THE

ECOLOGY

BOOK

TOM HENNIGAN
& JEAN LIGHTNER

Master Books®

First printing: March 2013
Fourth printing: June 2021

Master Books®
P.O. Box 726
Green Forest, AR 72638

Master Books® is a division of the
New Leaf Publishing Group, Inc.

ISBN: 978-0-89051-701-7
ISBN: 978-1-61458-317-2 (digital)
Library of Congress Number: 2012955728

Cover by Jennifer Bauer

Unless otherwise noted, Scripture quotations are from the New King James Version of the Bible.

Please consider requesting that a copy of this volume be purchased by your local library system.

Printed in China

Please visit our website for other great titles:
www.masterbooks.com

For information regarding author interviews, please contact the publicity department at (870) 438-5288

Master
Books®
A Division of New Leaf Publishing Group
www.masterbooks.com

Table of Contents

Our Best-selling Wonders of Creation Series is getting even better!

The series is being developed with an enhanced educational format and integrated with a unique color-coded, multi-age level design to allow ease of teaching the content to three distinct grade levels.

BUILDING MEMORIES

How to use this book

The Ecology Book has been developed with three educational levels in mind. These can be utilized for the classroom, independent study, or homeschool setting. For best possible comprehension, it is recommended that every reader examine the text on the off-white background. More skilled readers can then proceed to the green sections as well. Finally, the most advanced readers may read through all three sections. Look for the following icons and special features throughout the book:

> **Level 1**
> **Level 2**
> **Level 3**

Level One

- Text on off-white background
- 5th to 6th grades
- The basic level is presented for younger readers and includes the Building Memories and Wonder Why segments.

WONDER WHY

WHO
WHAT
WHERE
WHY
HOW

THE BIG FIVE Each chapter addresses the who, what, where, why, and how of important ecological and historical information. This information may be found on any level.

Level Two

- Text on green background
- 7th to 8th grades
- This middle level delves deeper into ecological issues related to today's environmental concerns, utilizing Words to Know and practical applications.

Words to know These words found throughout the upcoming chapter present a quick glimpse of important concepts coming up.

Level Three

- Text on blue background
- 9th to 11th grades
- This upper level incorporates concepts and theories related to all subject matter included in the text, as well as unique information within the Dig Deeper sections.

Make a Difference!

Watch for ways you can make a difference in your home or community. It is a joy to give back when you can!

ECOLOGIST *from the pages of history*

Look in the pages of the history of ecology at the men who have influenced the world in the realm of ecology with these level three insights.

Chapter 1

Harmony in Creation

> **Level 1**
> **Level 2**
> **Level 3**

Many people struggle to understand the concept of the Trinity. To some it seems contradictory (confusing). They reason that if there are three persons, then there must be three gods. This is not the case. The Bible makes it very clear that there is only one true God.

Hear, O Israel: The LORD our God, the LORD is one. [echad in Hebrew] (Deut. 6:4).

It is possible for more than one person to be considered one.

Therefore a man shall leave his father and mother and be joined to his wife, and they shall become one [echad] flesh (Gen. 2:24).

Even in our culture today, we understand that one team is made up of more than one person. The fact that the Father, Son, and Holy Spirit are one has some important implications.

	THE BIG FIVE	
WHO		Who created the world?
WHAT		What does Trinity mean in relation to God?
WHERE		Where do we see relationships in life?
WHY		Why is an understanding of a triune God important for the origin of love and the observations that most organisms are in cooperative relations with one another?
HOW		How does the Bible provide the basis for science?

Words to know

Mutualism (symbiosis)
Commensalism
Parasitism
Trinity
Harmony
Hypotheses

Scientists classify relationships between different organisms into several general categories. Three of these are mutualism, where both organisms benefit from the relationship; commensalism, where one organism benefits while the other seems unaffected; and parasitism in which one partner takes nutrients from another, sometimes causing harm. Often organisms produce a combination of one or more of the above relations, and it is difficult to classify them.

God Reveals Himself as Triune

From the very first chapters of Genesis, God reveals who He is and how He created the world. In the original Hebrew language, the word for God (Elohim) is plural. Oddly, it is used with a singular verb. This would seem grammatically incorrect, something like saying "they creates" in English. Yet this is the pattern that is regularly seen in the Hebrew: God (Elohim) is plural, but the verb describing His action is singular. Bible scholars have often described this as the plural of majesty. In other words, the word for God is plural to indicate how great He is.

Within the first chapter of Genesis, it becomes apparent that more than this is going on.

Then God said, "Let Us make man in Our image, according to Our likeness; let them have dominion over the fish of the sea, over the birds of the air, and over the cattle, over all the earth and over every creeping thing that creeps on the earth." (Gen. 1:26).

Notice the first three pronouns that are used (us, our, our). Who is God talking to? Himself? Some angelic beings? The answer is found in the next verse.

So God created man in His own image; in the image of God He created him; male and female He created them. (Gen. 1:27).

These verses provide us with the first clue that God is triune. While the word Trinity does not appear in the Bible, the concept does — beginning in Genesis. The teaching of trinity has been important throughout Church history. It can be summed up as the belief that there is one God who has always existed as three persons: Father, Son, and Holy Spirit.

Harmony within the Godhead Reflected in Creation

God exists as a Trinity, so He is, by nature, relational. This makes sense of the fact that God is love (1 John 4:16) and God is one (Deut. 6:4). God is love and has always been so. There has always been loving harmony between Father, Son, and Holy Spirit, even before creation. God is one, and He desires for people, who were created in His image, to be able to share in this, just as we find in Jesus' prayer for his disciples:

I do not pray for these alone, but also for those who will believe in Me through their word; that they all may be one, as You, Father, are in Me, and I in You; that they also may be one in Us, that the world may believe that You sent Me. (John 17:20–21).

The implications of the Trinity extend beyond this. All of creation was originally created with important harmonious relationships. Plants were created as food for animals and people (Gen. 1:29–30). People were created to rule over and care for creation (Gen. 1:25–28; 2:15). As we explore the world around us, we will find many other examples of relationships. In this book, we will emphasize the many harmonious and mutual symbiotic relationships still in effect today.

Harmony Broken

When God finished creating, He declared everything "very good" (Gen. 1:31). Unfortunately, this did not last for long. People were created with a will and the ability to make choices. Soon Adam and Eve made the choice to disobey the one command that God gave them. They ate from the tree of the knowledge of good and evil. In doing this, they came to experience brokenness and death. There was brokenness in their relationship with God, and they hid from Him (Gen. 3:8). There was brokenness in their relationship with each other (Gen. 3:16), and there was brokenness in their relationship with the world around them (Gen. 3:17–19; Rom. 8:20–22).

Natural Disharmony

Because of brokenness and death, disharmony is also found in creation. Because the Bible describes the creation as a very good creation in the beginning, creation biologists recognize that disharmony is a result of relationships gone awry since the Fall (Gen. 3). They seek to explain how they may have come about. Hypotheses (educated guesses) include negative genetic changes of the original good organism and/or the movement or displacement of the creature from where it originally performed its good function.

Reflections

- Can you think of a relationship in nature that is positive for everyone and may reflect how things were originally created to be?
- Can you think of a relationship in nature that reflects disharmony associated with the Fall?

Chapter 2

Ecology: The Study of "Our House"

The darkness surrounded us as we inched down the woodland trail. As the nearly full moon rose above the horizon, I could barely see the teens behind me. The science students huddled together for protection because they had never walked in a forest at night.

You might ask, "Why would anyone do that?" Well, the students wanted to experience the creatures and overcome their fear of the dark. The forest at night is filled with mystery and beauty, and as we continued farther, wonder replaced their fear. Above our heads, baby owls screamed for their mamas. Flying over the wet grass, fireflies flashed their lanterns. Across the lake, bullfrogs bellowed their "jug-o-rum" chorus. It was an experience the students would never forget, and they had lots of questions. Does the kind of water affect where bullfrogs live? What is a firefly's light used for? Do owls have relationships with other animals? How is the soil dependent on the forest and the forest dependent on the soil? Why do some creatures come out at night and others come out in the day? These are the sorts of questions ecologists are interested in. An ecologist studies the relationships creatures have with each other and with their habitats (home).

> Level 1
> Level 2
> Level 3

THE BIG FIVE

WHO	Who were William Derham and Ernst Haeckel?
WHAT	What does the word ecology mean?
WHERE	Where do ecologists primarily work?
WHY	Why do natural cycles, such as the water cycle, exist?
HOW	How were scientific names for animals created?

Words to know

Abiotic	Evaporation
Baraminology	Evapotranspiration
Biodiversity	Hybridization
Biome	(hybrids)
Biosphere	Population
Biotic	Precipitation
Condensation	Protocol
Discharge	Recharge
Ecology	Species
Ecological System	Sublimation
(Ecosystem)	Watershed

BUILDING MEMORIES

As a family, research the sounds that are made by some of the nighttime animals in your area. Many can either be googled from your home computer or found at your local nature center, bookstore, or library. Find a natural location you are familiar with. Travel to a secluded spot just before nightfall, and together, listen to the sounds of the night. It may take several minutes of pure quiet before hearing the darkness come alive, but it will be worth it. Fear, at its basis, is often a product of the unknown. By researching ahead of time, you can learn much, and the Holy Spirit of God can dispel other fears if you let Him. The relationships between the creatures of the night are just as amazing as those of the day. God's encouraging words to His son Joshua may apply here: "…*Be strong and of good courage; do not be afraid, nor be dismayed, for the* LORD *your God is with you wherever you go.*" (Josh. 1:9).

The Study of Ecology

Ecology is the study of intricate relationships between biotic (living) communities and their abiotic (nonliving) environments. The biotic community consists of species and populations. The species concept is complicated and controversial, but most ecologists would agree that, at its basic level, a species is a group of creatures that can reproduce. The offspring of these parents are then able to reproduce with each other. For example, the baby owls we heard on the night hike were barred owls (*Strix varia*). Another species in this genus is the spotted owl (*Strix occidentalis*). They have been considered separate species because barred owls were thought to produce fertile offspring only with barred owls and spotted owls with spotted owls. However, one reason the species concept is so difficult is that different "species" can often mate with each other and produce fertile young. This is the case with the barred and spotted owls. The barred owl is increasing in the western habitats of the spotted owl, and hybridization (sexual crossing between the two species) has been reported. These hybrids are producing fertile young.

The barred owl, *Strix varia*, (left) is typical to North America and is fairly common. It is able to breed with the less common spotted owl, *Strix occidentalis* (right).

Though the species concept is helpful to creation ecologists, it is not enough. Biblical creationists are currently interested in defining and identifying the "kind" God spoke about in Genesis. This has led to a unique creationist area of biology called baraminology. Baraminology comes from the Hebrew— *bara* [to create] and *mín* [kind]—and means the study of the created kinds. Because the ability to hybridize means that two creatures are very similar in their behavior and design, creation biologists infer that the barred and spotted owls came from the same owl ancestor that flew off the Ark 4,500 years ago. Since we believe that God wants His creatures to persist in this fallen world, we are very interested in how they were designed to respond to and survive in changing environments. We believe some of these answers will help us get to the reasons why we see an amazing amount of variation in animals like owls, frogs, and fireflies.

What Is a Population?

The population concept is also difficult to define. But it can be described as the group of one species in an area that have equal chances of mating with one another. For example, if there were barred owls in one area and they never interacted with barred owls two miles away, they would be considered two distinct barred owl populations because no mating was taking place between them. Most of the time creature relationships and interactions are much more complicated than that, and defining populations becomes more of a problem.

WONDER WHY

Why do animals have such funny scientific names?

The reason these owl names look so strange is because they are written in Latin. Other creatures may have names written in Latin and/or Greek. Writing these names in Latin or Greek allows ecologists to be consistent in the naming of organisms and speak the same language with one another. *Strix* (Latin for a mythical night bird) is their genus name and refers to the group of owls that do not have little ear-like feathers on their heads. The second name, *varia* (Latin for various) or *occidentalis* (Latin for Western), represents their species names and is written in lowercase. This second name refers to the specific creature we are talking about and is often based on an unusual trait or the location in which it is found. When typed both words are italicized.

ECOLOGIST
from the pages of history

Ernst Haeckel

Ernst Haeckel (1834–1919) was a German zoologist and naturalist who invented the name ecology (oecologie). The word comes from the Greek *oikos*, which means "the house" or "place to live," and *ology* means "study of." The word gradually changed from oecologie to ecology.

Haeckel was an outspoken evolutionist and held the false belief that creatures repeated their evolutionary history during their embryonic development. He wanted nothing to do with God or the Bible. Haeckel greatly exaggerated similarities in his embryo drawings to promote his ideas about evolution. When embryos from different animals are compared in early development, they look quite different. Toward the middle of their development, there are more similarities for a brief time. Then as they move toward full development, they look very different once again. Unfortunately, Haeckel only saw what he wanted to see. He went to the grave believing falsehoods and rejecting his Creator. May we always trust in the Word of our Creator, for it continues to be a lamp for our feet and a light to our path (Ps. 119:105).

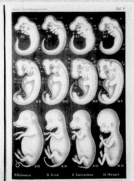

False chart of embryonic development

What Exactly Is an Ecosystem?

The ecological system or ecosystem is the location where organism populations that make up the biotic community interact with each other and with their environment. The ecosystem concept was developed by British ecologist George Tansley in the 1930s and can be as small as a drop of water or as large as a biome. Biomes are large ecosystems that include familiar places like the desert, tundra, deciduous forest, or tropical rainforest. The global sum of all the biomes of the world is the biosphere.

The size of the ecosystem of interest is often determined by the questions ecologists are interested in. The more we study these systems, the more we realize that relationships drive them. For example, it is the intricate relationships that organisms have with one another and their surroundings that enable the biosphere to regulate itself. Everything about these systems speaks of design, and to try to explain their existence as products of random, accidental processes goes against what the Bible teaches and our own experiences. Isaiah 45:18 says:

For thus says the LORD,
Who created the heavens,
Who is God,
Who formed the earth and made it,
Who has established it,
Who did not create it in vain,
Who formed it to be inhabited:
"I am the LORD, and there is no other."

The amazing differences between species, genes, and ecosystems are staggering to ecologists, yet their interactions often produce a well-organized system and together describe biodiversity. Everything is either directly or indirectly connected to everything else, and everybody seems to have a job. Some creatures are taking raw materials that are being cycled through the system and producing products others need. Other organisms are transporting useable products and raw materials somewhere else. Still others are making toxins harmless and recycling them into useable forms. These jobs often require the cooperation of two or more completely different creatures. This is why biodiversity is an important concept in ecology. Ecologists must consider all of the organisms when managing for an ecosystem's health. Understanding all of these complicated relationships makes our technology rather pale in comparison. The Bible teaches that the earth and its systems were designed by a God who is an Engineer without equal. He has made us, in part, to understand it and take care of it. The rest of this book will highlight just a few of these breathtakingly complex partnerships.

ECOLOGIST
from the pages of history

William Derham

William Derham (1657–1735) was a Christian pastor who had a great interest in the sciences, including the study of birds, insects, weather, and stars. He is known for writing a large work called *Physico-Theology*, which argued strongly for design in nature by the God of the Bible. He was among the first to write about the study of science as a stewardship from God, much like the emphasis of this book. In some senses he could be considered a father of modern ecology.

Where Do Ecologists Work?

It is safe to say that most ecologists work in the wild places of the planet. They may follow bears through swamps, hang off tropical trees, dig in forest soil, or study water creatures. Measurements like types of plants, temperature differences, rain or snow amounts, chemicals traveling through the environment, and how organisms are related to each other are important in the study of ecosystems. For most ecologists, the work continues in a laboratory, where creatures may be watched under a microscope, chemical samples tested, animal behaviors compared, and mathematical calculations done in order to make sense of their measurements.

Hydrologic Cycle: The Watering Down of an Ecosystem

God's Word has a lot to say about the importance of water. Earth's water is always moving and is a crucial abiotic player in an ecosystem. How often do we take it for granted? When was the last time we thanked God for a rainy day rather than letting it cloud our mood? All God's creatures are dependent on it. It moves over the ground, under the land, and through the air. The water of earth has been used and reused since the beginning and has no beginning or end to its cycle. Think about it. God has designed this cycle in such a way that water is continually cleaned for drinking and reuse.

1 How does this cycle work? Let us begin with a large body of water like a giant lake or the ocean.

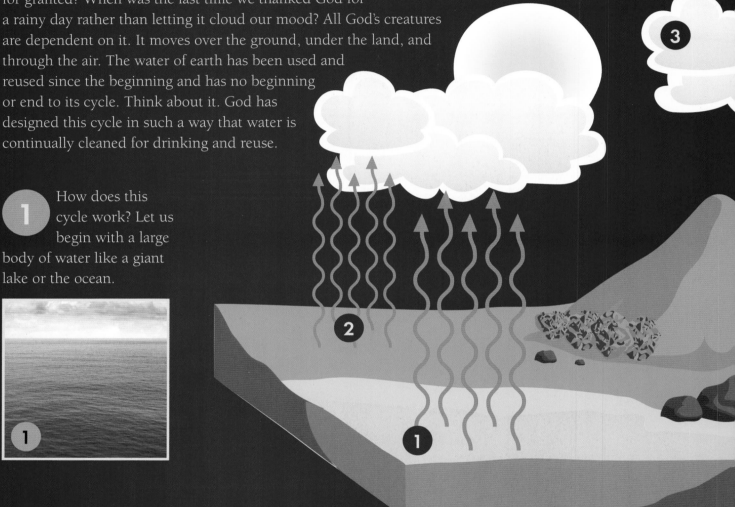

2 The warmth from the sun drives this cycle by evaporating the surface water and changing it from a liquid to a vapor (gas). In this manner, any solids from salts and other minerals, along with pollutants, are left behind. The water is now purified and rises in the atmosphere by air currents. Under conditions of ice and snow, the solid phase of water can directly change to its gas phase, skipping the liquid phase altogether. This process is known as sublimation. Trees and other plants also contribute evaporated water to the atmosphere through the process of evapotranspiration. As plants draw water from the ground, water travels into the roots, up the stems, and into the leaves, where it evaporates off the leaf surfaces into the atmosphere. Other organisms, including you, produce water as a waste gas when exhaling.

You care for the land and water it; you enrich it abundantly. The streams of God are filled with water to provide the people with grain, for so you have ordained it. You drench its furrows and level its ridges; you soften it with showers and bless its crops. You crown the year with your bounty, and your carts overflow with abundance. The grasslands of the wilderness overflow; the hills are clothed with gladness. The meadows are covered with flocks and the valleys are mantled with grain; they shout for joy and sing…. He makes springs pour water into the ravines; it flows between the mountains. They give water to all the beasts of the field; the wild donkeys quench their thirst. The birds of the sky nest by the waters; they sing among the branches. He waters the mountains from his upper chambers; the land is satisfied by the fruit of his work. He makes grass grow for the cattle, and plants for people to cultivate bringing forth food from the earth: wine that gladdens human hearts, oil to make their faces shine, and bread that sustains their hearts (Ps. 65:9–13; 104:10–15 NIV).

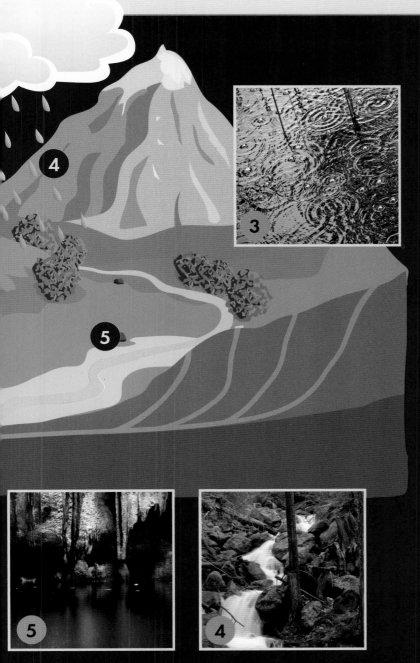

3 Generally, the higher you go in the lower atmosphere, the colder it gets. As the rising water vapor from all of the above sources cools, it changes back into a liquid through the process of condensation. Water molecules condense on solid particles like pollen, bacteria, dust, and various pollutants to form clouds. They collide with each other, grow, and eventually drop out of the sky as precipitation because the air can no longer hold their increasing weight. Depending on conditions, the precipitation can be in the form of rain, snow, sleet, or hail.

4 The returning water recharges the planet with clean water either directly into the oceans and other water bodies or indirectly as surface water runoff from the land.

5 Fresh water often becomes stored in important places of the biosphere that include ice caps, under the ground surface in aquifers where water saturates the soil and rock, and in freshwater lakes. Water access and availability determines creature design, biodiversity, and symbiotic relationships.

The water cycle is one of many natural cycles that were clearly designed by our loving supernatural God. When we recognize how essential it is for life on earth, it should encourage us to thank and trust our Creator

> "...I mean only that a wrong attitude toward nature implies, somewhere, a wrong attitude towards God..."
> — T.S. Eliot

Chapter 3

> Level 1
> Level 2
> Level 3

The Dominion Mandate

Did you know that some people blame Christians for the bad things that have been done to our earthly habitat? Some think that the Bible teaches that it is okay for people to dirty the water or kill living things at any time for any reason. This is not true. The truth is that God blessed Adam and Eve in Genesis 1:28 and gave this earth and all of His creatures as gifts to be cared for. God wants us to guard and protect His creation by being good stewards. This means that we love one another and that we wisely care for the land and its creatures. What is very special about all of this is that God wants us, with the help of His Holy Spirit, to be like Him. Yes, people from all walks of life have done horrible damage to God's creatures and their habitats. However, this behavior is not encouraged in the Bible. Rather, this behavior is a result of those who walk in sin and who are in disobedience to God. Are you up to the challenge? Will you be one of God's guardian stewards?

THE BIG FIVE

WHO	Who was given the dominion mandate?
WHAT	What is citizen science?
WHERE	Where can I find information about helping with real scientific research in ecology?
WHY	Why is an understanding of the dominion mandate crucial to biblical stewardship?
HOW	How is biblical stewardship an important component of our relationship with Christ?

Words to know

Dominion mandate
Stewardship
Organism
Protocols

What Is Biblical Stewardship?

In Genesis 1:28, God said to them, *"Be fruitful and multiply; fill the earth and subdue it; have dominion over the fish of the sea, over the birds of the air, and over every living thing that moves on the earth."*

If God made us to rule, what does that mean? What does this rulership look like? God's command to have dominion and subdue creation has been misunderstood. It is true that people have used this verse as a justification to harm many ecosystems on the planet. However, this is not what it teaches. The Hebrew for having dominion and subduing carries the meaning of being in charge. The word steward comes from the Old English word *stigweard*, which means "guard of the hall." The word implies that a steward is responsible for taking care of something for someone else.

We are created in the image of our Maker, and He has designed us to be the caretakers and protectors of His creation. When we put this all together — ruling, subduing, having dominion, and stewarding — it is like a forester who uses an ecosystem approach to a timber stand. An ecosystem approach means that the forester uses her knowledge about the relationships that make a forest ecosystem what it is. She takes charge of the trees in a way that they can be used as needed resources to help people. At the same time, trees are taken in a way that the forest will continue to produce new trees for generations to come. In this manner, the forest steward exercises a kindly and charitable dominion that balances the use of timber for market while maintaining the important and healthy relationships between the organisms, their environment, and the timber. The godly forester is devoted to providing needed resources to people while caring for God's creation. Should not the church be at the forefront of these worthy activities? Subduing and ruling means that we have the responsibility of bringing increased order, vitality, fruitfulness and diversity to God's earth.

19

The Shepherd King Analogy

Throughout Scripture, God gives us beautiful imagery about Himself as our Shepherd, Ruler, and King. He is our model of what true stewardship and dominion look like.

Life for a shepherd was challenging. Hours became days and seasons changed as he braved the elements, keeping a watchful eye on his sheep. It was a life of devotion and self-sacrifice to care for the needs of his vulnerable animals. It took courage and patience, vigilance and compassion.

David, the boy shepherd and psalmist who became Israel's king, wrote:

"The LORD is my shepherd; I shall not want. He makes me to lie down in green pastures; He leads me beside the still waters. He restores my soul; He leads me in the paths of righteousness For His name's sake. Yea, though I walk through the valley of the shadow of death, I will fear no evil; For You are with me; Your rod and Your staff, they comfort me." (Ps. 23:1–4).

The shepherd ruler is a servant ruler. He leads in wisdom so his sheep are guided, refreshed, and encouraged.

When danger comes, the lives of his sheep come first, ahead of his own. His staff was a short wooden club, which had a lump of wood on the end, often filled with sharp objects. It was the weapon he used to defend himself and the sheep. The rod was like a shepherd's crook and was used to catch and pull back any sheep that strayed from the safe path.

At the end of the day, as the sheep were gathered into the fold, the shepherd would hold the rod across the entrance, low to the ground, and every sheep had to pass under it (Ezek. 20:37; Lev. 27:32). This was done so that each sheep could be examined carefully for any injuries it might have received that day. Then the shepherd would tend to them as needed. His rod and staff were truly for their safety and protection, and for those reasons, they brought great security

Jesus is the Shepherd of our souls.

I am the good shepherd. The good shepherd gives His life for the sheep. But a hireling, he who is not the

shepherd, one
who does not
own the sheep, sees
the wolf coming
and leaves the
sheep and flees;
and the wolf catches the
sheep and scatters them. The hireling
flees because he is a hireling and does not
care about the sheep. I am the good shepherd; and I
know My sheep, and am known by My own. As
the Father knows Me, even so I know the
Father; and I lay down My life for the
sheep (John 10:11–18).

In those days, the shepherd would
lead the flock to make sure that the path
was safe, and then the flock would follow. It is
absolutely true that the sheep knew the shepherd's
voice and would not respond to a stranger. May we
always follow our Shepherd's voice and steward His
creation as He stewards us.

Reflecting God's Image

Human life is precious, and we were meant to guard it because we are the only beings made in the image and likeness of God (Gen. 1:26–27; James 3:9). Though the meaning of the "image of God" has been discussed and argued for a very long time, at the very least it means that God made us His representatives on earth. It also might have to do with our character and conduct. He gave us spirits that have been made to live forever. God is Spirit, and we must worship Him in Spirit and in truth (John 4:24). Maybe we have the potential of having similar abilities to think, love, judge, and care for others. Whatever it means, to be made in His image is the reason we must respect and love one another and steward the planet wisely. This means we must protect unborn babies, older people, those who are helpless, and those who cannot speak for themselves (Ps. 82:3–4; Prov. 31:8–9; Isa. 1:17). We also must make sure that our habitats are clean and healthy and encourage others to do the same. After all, we were made so we could be caretakers of each other and the earth for God. Is it not an honor to be given this great responsibility by the Creator?

O Lord, our Lord, How excellent is Your name in all the earth, Who have set Your glory above the heavens! Out of the mouth of babes and nursing infants You have ordained strength, Because of Your enemies, That You may silence the enemy and the avenger. When I consider Your heavens, the work of Your fingers, The moon and the stars, which You have ordained, What is man that You are mindful of him, And the son of man that You visit him? For You have made him a little lower than the angels, And You have crowned him with glory and honor. You have made him to have dominion over the works of Your hands; You have put all things under his feet, All sheep and oxen—Even the beasts of the field, The birds of the air, And the fish of the sea That pass through the paths of the seas. O Lord, our Lord, How excellent is Your name in all the earth!

Psalm 8

Be prepared!

Eating snacks and drinking sodas in the car is fun, but don't toss those wrappers and bottles out the window! Trash can harm wildlife as well as create an unsightly mess!

Make sure you keep a small plastic bag in the car at all times to serve as a mobile trashcan when you need it. Or take two, one for recyclables and the other for throwaway items.

Ecology in Practice: Getting Involved

As an ecology student, there are many opportunities to participate in citizen science projects where you can study specific creatures and/or ecosystems in your area and help real wildlife ecologists do research. Good science involves careful thinking, making good observations, using equipment properly, collecting important data, and following specific directions as to how to collect and record the data. These specific directions are called scientific methods, and each citizen science project will have specific protocols (ways of doing things) to follow. These protocols are important because they allow all researchers to be consistent in collecting data. If different people collected data in any way they wanted, it could affect their understanding of what is really going on in the ecosystem. Wrong conclusions could cause ecologists to steward creation badly, and this must be avoided. Remember that you might be handling these creatures, so do not forget, handle them with care, for they are the Lord's. If you want to learn more about involvement in these types of programs, do an online search for the projects below using the name of the program and the phrase "citizen science."

Monarch Butterfly Research

Bird Monitoring

North American Amphibian Research and Monitoring

Black Bear Education and Hibernation Data

Lady Bug Monitoring

Stream Monitoring and Ecology

> "Nothing would be more tiresome than eating and drinking if God
had not made them a pleasure as well as a necessity."
— Voltaire

Chapter 4

What's on the Menu Today?

> Level 1
> Level 2
> Level 3

Did you know that in the beginning, God created us to eat only plants? He created the animals the same way.

And God said, "See, I have given you every herb that yields seed which is on the face of all the earth, and every tree whose fruit yields seed; to you it shall be for food. Also, to every beast of the earth, to every bird of the air, and to everything that creeps on the earth, in which there is life, I have given every green herb for food"; and it was so. Then God saw everything that He had made, and indeed it was very good. So the evening and the morning were the sixth day…. The LORD God planted a garden eastward in Eden, and there He put the man whom He had formed. And out of the ground the LORD God made every tree grow that is pleasant to the sight and good for food. The tree of life was also in the midst of the garden, and the tree of the knowledge of good and evil (Gen. 1:29–31; 2:8–9).

When God first created, there were no cats eating mice or owls eating rabbits. Things have sure changed, have they not? In this chapter, we will be talking about relationships between creatures that are eaten by some and those who eat others. As you read, think about how the eating has changed since Adam and Eve. Even more important, be happy about the fact that God will remake the earth so all relationships are good, just like He first planned. There will be no more death and no more tears, and that is a promise (Rev. 21 and 22)!

Words to know

Abiotic	Habitat
Autotroph	Herbivore
Biotic	Heterotroph
Biodiversity	Niche
Carnivore	Omnivore
Chemoautotroph	Photoautotroph
Decomposer	Photosynthesis
Ecological pyramid	Saprophyte
Food chain	Scavenger
Food web	Trophic level

24

Why do we kill and eat animals?

God currently upholds the world in a way that includes meat eating. Eventually sin will be completely dealt with, and the world will return to its original harmony. Though God has given us permission to eat meat, it is not how things were meant to be. Consider this wonderful promise from God:

The wolf also shall dwell with the lamb, The leopard shall lie down with the young goat, The calf and the young lion and the fatling together; And a little child shall lead them. The cow and the bear shall graze; Their young ones shall lie down together; And the lion shall eat straw like the ox. The nursing child shall play by the cobra's hole, And the weaned child shall put his hand in the viper's den. They shall not hurt nor destroy in all My holy mountain, For the earth shall be full of the knowledge of the LORD *As the waters cover the sea* (Isa. 11:6–9).

THE BIG FIVE	
WHO	Who was Eugene Odum?
WHAT	What is ecosystem ecology?
WHERE	Where does energy go?
WHY	Why do many of God's creatures have to die and be eaten?
HOW	How have relationships changed since creation?

Balance and Biodiversity

Despite the fact that we live in a fallen world, both biblical and secular ecologists are interested in how energy, in the form of light and food, moves through and affects the health and biodiversity of ecosystems. Biodiversity refers to the number and type of organisms and how they are related. Those who measure biodiversity are also interested in how many different ecosystems are present. The study of how this energy moves through the living and nonliving parts of the environment is called ecosystem ecology.

Each organism has one or more habitats (homes) that it prefers. Just like your home, the habitat consists of shelter, food, water, space, and how the space is arranged. Not only do all God's creatures need a good habitat if they are to survive, but they also have an important function or niche. You can look at the niche concept in many ways. It may describe the type of habitats organisms need for their survival. Abiotic factors like what types of temperatures are needed, how much water is required, or types of soil nutrients available are important things that determine the habitat niche of a creature, bunchberries being a good example. You can also view the niche as how organisms go about finding shelter or food or how they meet their reproductive needs. Finally, you can think of the niche as the kind of job a creature has. For example, is it an important food for others? Does it keep certain organisms from overpopulating? Does it make a good shelter? Notice that these concepts are all based on the study of relationships.

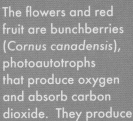

The flowers and red fruit are bunchberries (*Cornus canadensis*), photoautotrophs that produce oxygen and absorb carbon dioxide. They produce a red drupe (fleshy fruit containing a single seed) that is a food source for moose and birds. It is found in northern North America and the mountains of the south.

The American toad (*Anaxyrus (Bufo) americanus*), changes as it goes through life stages of metamorphosis. As tadpoles, they are herbivores and eat algae growing underwater. When they become adults, they eat a variety of things, including insects, earthworms, slugs, and snails. If you watch closely, you will see their eyes close and flatten against their head during swallowing. The underside of the eye is actually helping push food down the esophagus. One American toad can eat up to 1,000 insects every day and is important in keeping insect populations down. They are also important food sources for snakes like the hognose snake (*Heterodon platirhinos*) and the garter snake (*Thamnophis sirtalis*). Toads get water, not by drinking, but by absorbing it through their skin.

Energy, Ecosystems, and Ecologists

Energy is the power to do work. Energy allows living things to move, grow, and do things. Plants can use energy from the sun to grow, develop, and do things. Other creatures like dogs, deer, fungi, and birds get their energy from food so they can grow, develop, and do things. Ecologists can study how energy flows through an ecosystem by studying who eats whom.

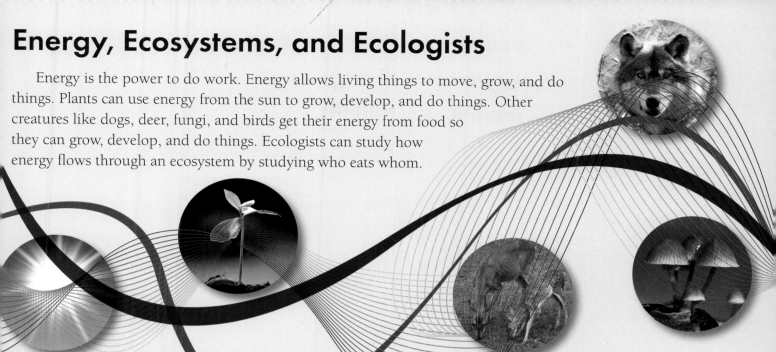

Autotrophs or Heterotrophs?

Photoautotroph

Autotrophs are organisms that get their carbon from carbon dioxide (CO_2). This carbon will be used to build organic compounds like sugar and protein. Phototrophs get their energy for life from light. Chemotrophs get their energy from chemicals in their environment such as hydrogen and sulphur compounds. They all are called producers. These creatures are the foundation for the biotic community because everyone else is dependent on them. Heterotrophs, also called chemoorganotrophs, are consumers and must obtain both their energy and carbon by eating other organisms. Heterotrophs include people, fungi, and most animals. If photoautotrophs and chemoautotrophs did not exist, there would be no chemoorganotrophs.

There are different kinds of heterotrophs, and they include carnivores (meat eaters), herbivores (plant eaters), omnivores (meat and plant eaters), decomposers (recyclers of nutrients from dead organisms, putting them back into the soil), and saprophytes (organisms that grow on, and get nutrition from, dead organisms or parts of organisms). Saprophytes may sometimes be decomposers when they help to recycle important nutrients.

Herbivore

Carnivore

Omnivore

Decomposer

Parasite

Energy and Ecosystems

Scientists use models to help them understand science concepts. Models can be diagrams, computer programs, or math equations. Food chain and food web diagrams are helpful models to show how energy flows in an ecosystem. Each step in a food chain or web is called a trophic level. For all ecosystems, the producers are the first trophic level. Another model used to show the flow of energy through an ecosystem is an ecological pyramid, such as the one above. These models show the relative amounts of energy for each trophic level but can also be designed to show the relative number of organisms. As you move from the bottom to the top, there is less energy available for the next trophic level, and the number of organisms decreases. That means that in a particular ecosystem, there will be much more grass than there are foxes.

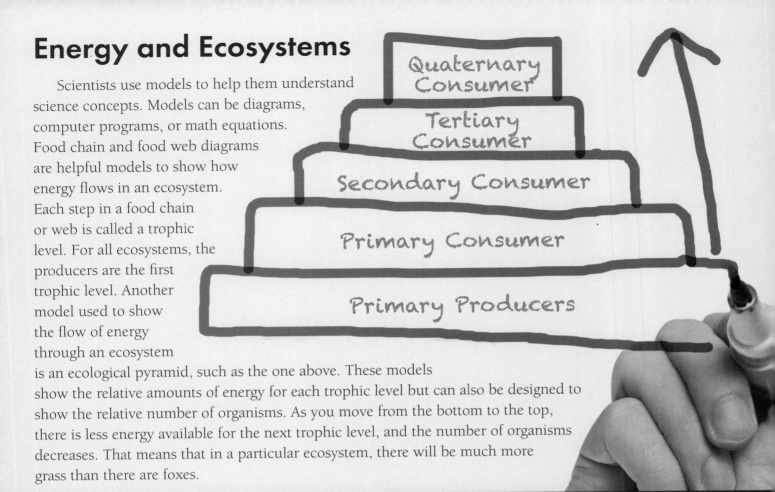

Quaternary Consumer

Tertiary Consumer

Secondary Consumer

Primary Consumer

Primary Producers

Photoautotrophs

Chemoautotrophs

How Organisms Get Energy

There are two basic types of autotrophs: photoautotrophs and chemoautotrophs. Photoautotrophs, like plants, algae, and cyanobacteria, use light energy from the sun and carbon from CO_2 in order to make important carbohydrate nutrients like sugar. The process they use to do this is called photosynthesis. Photo means "light" and synthesis means to "build" or "make". The word refers to the idea that plants and other photoautotrophs can build important nutrients using light energy and CO_2. However, there are other fascinating ecosystems, like the deep oceans, where light does not penetrate. In those places, there are certain bacteria that can get their energy from chemicals such as sulphur compounds and their carbon from CO_2 to produce important carbohydrate nutrients. These chemoautotrophs are the producers in these environments and are the basis for the food web.

Heterotrophs (chemoorganotrophs) cannot directly use the energy from light or sulphur compounds, nor can they use the carbon from CO_2 and must ingest other organisms. When heterotrophs ingest food, the digestive system breaks it down and energy is released and carbon is obtained. That energy and carbon are then used to build important compounds for life.

Links in the Food Chain

  Plants, like grass, can use the energy from the sun to make food. An animal, like a grasshopper, may get its energy by eating the grass, a blue jay may get its energy by eating the grasshopper, and a fox may get its energy by eating the blue jay. The above example of how energy moves from sun to grass to blue jay to fox is called a food chain. When ecologists study many food chains put together, they are studying a food web.

Producers, Consumers, and Decomposers

  The first trophic level represents the primary producers (grass), followed by the primary consumers or herbivores. After the herbivores come the secondary consumers, tertiary consumers, quaternary consumers, and decomposers. The numbers of trophic levels vary with the type of ecosystem.

  You might ask: Where does the energy go when it is used? Energy cannot be created or destroyed by natural processes; it just changes from one form into another. This is based on a solid principle of physics called the law of conservation of energy. This law says that the total amount of energy in a system (like an ecosystem) remains the same or is conserved over time. As you go from one trophic level to the next, the energy does not disappear; it just changes form into mostly unusable heat energy. Therefore, in the real world, the reason why there is more grass than foxes is because there is much less energy available to support a fox population than there is a grass population.

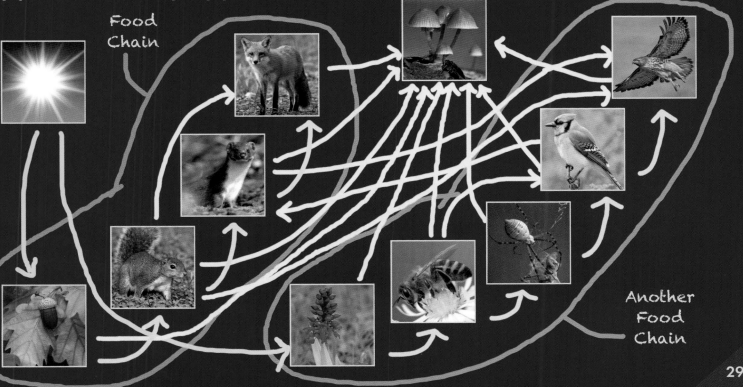

Sun Power

We know that plants need sunlight in a process used to make food, but we also need the sun. Too much sun can be bad for us, but a little bit can make all the difference in the world. Ever wonder why you get the "blahs" in winter? Your body may be needing sunlight since you are staying indoors on cold and cloudy days. When exposed to sunlight, our bodies produce vitamin D, which the liver converts into a substance the kidneys use for bone growth and maintenance. Get outside and soak up some sun soon!

Eugene Odum

American ecologist Eugene Odum (1913–2002) was one of the first to look at ecology from an ecosystem perspective by studying the interactions between the biotic (living) community and the abiotic (nonliving) physical aspects of the environment. He was concerned with how energy and nutrients flowed through ecological communities and how these affected environmental health. He studied the ecosystem as a series of interconnected living and nonliving relationships, a concept not well received by the majority view of the day but well supported by current research. In 2007, the University of Georgia began the Odum School of Ecology, which is the first stand alone research wing dedicated solely to ecological research.

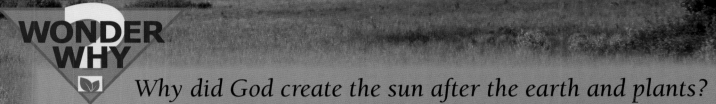

Why did God create the sun after the earth and plants?

Many people question the book of Genesis because if the sun drives ecosystems and the plants were created before the sun, that does not match what we observe, so they assume the Bible is wrong. First of all, remember that the Creator was an eyewitness to His creation and has briefly reported what took place. Light was the first thing that was created, and that is all plants need to carry on photosynthesis. On Day 4, He simply created a physical body from which light came. When you think about how secular scientists believe that it is the sun that we are solely dependent on and when you think about all the cultures that have worshipped the sun, you have to wonder. When God told us He made the sun on the fourth day, was He trying to communicate something? Was it to remind us that He is the source of life, not the sun? He is the Supreme Creator, the Author of life, and the only one upon which we must depend.

BUILDING MEMORIES

As a family, find a natural area with trails, and pack a picnic lunch or spend a night or two camping. Take note of the creatures you see. Look carefully under rocks, on plants, in the stream, and all over (See Appendix B). Observe the different habitats and the relationships the creatures have with each other and their environment. Take notes, draw pictures, and/or snap photos. Have a devotional time together, meditating on the following Scriptures: Genesis 1–3; 9, Acts 10, Romans 14, Revelation 21–22, and Isaiah 11:6–9, 65:25, in that order. Discuss how God originally created the earth to be, how it is today, and His promise of restored relationships in the future. Emphasize how this information is important to your lives and what you can do as a family to get closer to Christ, to one another, to your community, and to creation.

Chapter 5

> Lichens are fungi that have discovered agriculture.
> — Lichenologist Trevor Goward

> Level 1
> Level 2
> Level 3

Taking a Liking to Lichens

Lichens (pronounced *like-uns*) are all around us and are studied by scientists called *lichenologists*. Lichens are made of two or three creatures that live together in *symbiosis*. They look and act like plants but are not plants.

As the story goes, Freddy Fungus was strolling down the lane one day when he met a girl by the name of Alice Algae. At first sight, they took a likun to one another. Since Freddy is a good builder and an expert at getting water, he protects Alice and makes sure she isn't thirsty. Alice, being the good cook, provides the food, and together their marriage can be a long-lasting relationship where both benefit from each other's gifts.

THE BIG FIVE

WHO	Who was Simon Schwendener?
WHAT	What is a lichen?
WHERE	Where do lichens grow?
WHY	Why are lichens important to the creation?
HOW	How do lichens form symbiotic relationships?

Words to know

Asexual reproduction	Mycobiont
Bioindicator	Photobiont
Biomonitoring	Photomorph
Fragmentation	Soredia
Hypha (plural hyphae)	Spore
Lichen	Symbiont
Lichenologist	Thallus
Mycelium	

Looking Closer at Lichens

As silly and simplistic as the Freddy Fungus/Alice Algae story is, it may help you remember that a lichen is made of more than one creature. This relationship is a very complicated symbiosis between two or three totally different organisms: a fungus with an alga, a fungus with a cyanobacterium, or a fungus with both an alga and a cyanobacterium. The alga (plural algae) is an organism that can photosynthesize or produce sugars using the light energy of the sun. A cyanobacterium is a unicellular bacterium (plural bacteria) with a blue-green pigment that allows it to photosynthesize. The fungus (plural fungi) is a creature that does not photosynthesize and needs to get its nutrition from the environment. In fact, the lichen itself is a mini ecosystem that is an important habitat for yet other creatures living in its midst.

Letharia vulpina

Pseudevernia furfuracea

Divine Design

Lichens have also been designed to produce more than 500 unique chemical compounds that perform functions like repelling herbivores, killing microscopic parasites, preventing plant growth, and controlling the amount of light exposure.

Lichens, Lichens, Everywhere Lichens!

They live together in many places on earth, including very hard-to-live places. They can be growing on sidewalks, stained glass windows, old cemeteries, and buildings. There are lichens that are metal loving, and they can grow where rusty water drips on them, like under barbed wire fences. You can also find them in very cold or hot places like the North and South Poles, and on top of mountains or in hot deserts.

UV Protection

Jesus created lichens to be able to adapt (adjust) to changing and extreme environments. For example, just like you, lichens need to protect their cells from ultraviolet (UV) radiation damage. The difference is that while you have to go to the store and buy sunscreen, lichens were designed to produce an acid (usnic acid) that acts as a natural sunscreen to protect them against UV radiation. They seem to be able to adapt by changing the amount of this acid depending on how much UV radiation they are exposed to.

MAKE A DIFFERENCE
ECO-FRIENDLY

Off or On?

People assume it is always best to turn off lights when you are not in the room, but for some types of bulbs, if you will only be gone for a few moments, it is more energy efficient to leave them on. Visit http://energy.gov/ to get more information about different types of bulbs and energy efficiency tips you can use in your home!

How Does a Lichen Reproduce?

Lichens reproduce in two ways. One way is through sexual reproduction, though this is not very well understood. The fungus makes structures that produce spores. Spores are reproductive cells that can become a new individual. Spores will disperse from the lichen, and if one finds the right kind of environment it will produce a new fungus. A hypha from the baby fungus will grow, and if it finds the right photosynthetic partner, it will penetrate the future alga or cyanobacterium and produce a new lichen.

The second way they can reproduce is through asexual (no sex) reproduction. One type of asexual reproduction in lichens is called fragmentation. Fragmentation happens when a little piece of lichen breaks off and gets blown away to a new location. That little piece can grow into a brand-new lichen.

Fascinatingly, another asexual method for making new lichens comes in the form of little dustlike particles produced on lichen surfaces called soredia. Soredia are made of one or more algal cells surrounded by fungal hyphae. Wind can blow them great distances, and if they land in the right habitat, they can produce new lichens.

Lichen growing on dust. Each numbered tick on the scale represents a distance of .023 cm.

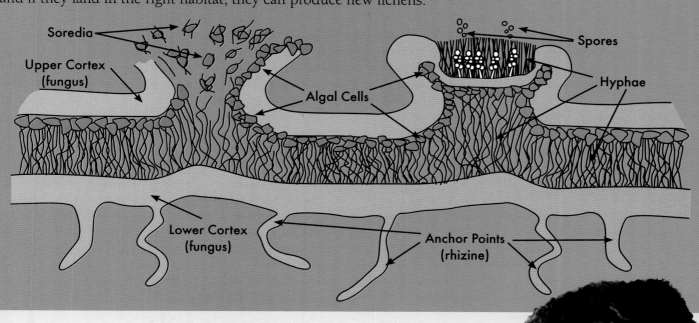

Soredia
Upper Cortex (fungus)
Spores
Algal Cells
Hyphae
Lower Cortex (fungus)
Anchor Points (rhizine)

Comparing Lichen Relationships with Human Ones

We, as followers of Jesus, were created to be in close relationship with one another. Our Father has given each of us gifts that are meant to be used to strengthen, protect, and encourage one another to live this life to its fullest and in obedience to Him. As you read about lichens, take time to study 1 Corinthians 12–14, and see if you can make comparisons about lichen and human relationships.

Lichens: A Complex Symbiosis

The fungal symbiont (partner) is called the mycobiont, and the photosynthetic partner (alga and/or cyanobacterium) is the photobiont. Though sometimes the creatures can live apart from one another, once they are in relationship, its thallus (lichen body) completely changes. When you compare what they look like by themselves and what they look like in relationship, you would not think it was the same creature. Consider these other wild and fascinating things to think about regarding lichens:

1. A fungus may change photobiont partners at different stages in its life.

2. Even though the same lichen grows in different geographical settings, it does not mean it always has the same photobionts.

3. Another fungus growing on a lichen may not only share the lichen alga but also exchange that alga for another one.

Interestingly, the form the lichen takes depends on its photobiont. Different forms of the same lichen are called photomorphs and why they do this is a mystery.

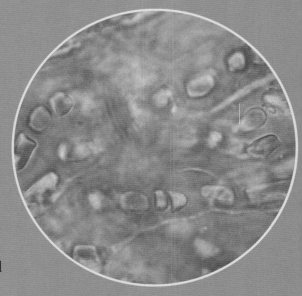

The cyanobacterium *Hyella caespitosa*, as observed within the lichen *Pyrenocollema halodytes*. Notice the fungal hyphae.

Choosing a Symbiont Partner

The fungus selects its symbiont partner, and many different types of fungi can be mycobionts. The greatest number comes from the ascomycete group of fungi, which have sac-like cells that produce spores. The names scientists give lichens are based on the fungus and not its photobiont. The fungus benefits its partner because it is better at finding and retaining water and other nutrients, forms the shape of the organism, and controls the reproduction.

The fungus produces long filaments known as hyphae (singular hypha), and large numbers of hyphae are called mycelia (singular mycelium) (C). Depending on the fungus, these hyphae may be involved with digesting and/or absorbing nutrients and taking part in reproduction. In lichens, they both surround and grow into the photobiont, which enables them to exchange nutrients. They also produce dense concentrations on the top of foliose (leafy) lichens that help keep other organisms out and reduce light intensity so photobiont cells do not get damaged.

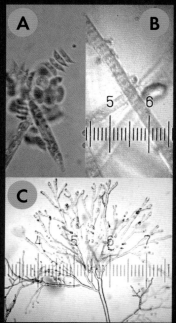

Photosynthetic Dependence

Recall that it is the photobiont that determines what the lichen looks like and where it can grow. There may be 23 genera and 100 species of lichen algae, and they mostly come from the genus *Trebouxia*, which are unicellular green algae (A). Cyanobacteria (B) come from the genus *Nostoc*. Since fungi cannot make crucial carbohydrate nutrients, like sugars, they depend on their photobiont to make these compounds by photosynthesis. Algae produce carbohydrates in the form of sugar alcohols, and cyanobacteria produce them in the form of glucose. An added advantage to the fungus who partners with cyanobacteria is that they can also obtain nitrogen compounds from them. Cyanobacteria can take atmospheric nitrogen and turn it into a form both plants and fungi can use. (See chapter 8 for the details of this amazing process.) The nitrogen compounds are crucial in making key chemicals such as DNA and proteins, without which all life would not be possible.

Shapes and Sizes

Lichens may have seven different shapes, but there are three shapes that are very common. **Crustose** (crusty) lichens look like someone might have painted on the rocks or sidewalks from which they grow. They can come in different colors and are tightly attached to what they grow on. **Foliose** (leafy) lichens look like leafy plants, and **fruticose** (shrubby/stringy) lichens look like little shrubs and are often branching, with round, stem-like structures. Some can also hang from trees and look like stringy moss.

Crustose Lichens

Foliose Lichens

Fruticose Lichens

Why are lichens so vital?

   Lichens are very important. They help make environments healthy by building soil so other plants and animals can come and live there. They prevent soil from washing away in some places. People have used them to make colors for clothes and medicines for colds. A lichen chemical, like usnic acid, is good at killing bacteria and is used as an antibiotic in some medicines. Some insects use lichen to blend into their environment, while others may eat some types as their major food source. Some birds and flying squirrels use lichens as an important material for their nests. Lichens are also important homes for other creatures, one of which might be the toughest creature on earth. Have you ever seen or heard of the water bear? They are also called tardigrades or moss bears.

Many birds use various lichens to build nests.

A healthy oak moss lichen (*Evernia prunastri*) from a non-polluted area. On the right is an *Evernia* affected by air pollution.

Air Quality and Biomonitoring

Lichens are important air-quality bioindicators. Bioindicators are living creatures that are used to determine the health of an environment. Biomonitoring means that we are studying areas where bioindicators live, and over time, we measure how changing environmental conditions affect them. Lichens are good to use for biomonitoring because they live all over the world, live for a long time, and absorb all kinds of chemicals from the environment. Lichens are different in their response to environmental factors, and how they respond depends on the species. Some are more sensitive to certain chemicals than others. Lichens have been monitored for air pollution for over 40 years, and if you are interested in taking part in studying air pollution effects on lichens, there may be a few projects around the world you can participate in from your own home.

The Growing Importance of Lichens

God's purposes for lichens can't be underestimated. They act as important carbon storage areas, regulating the chemical carbon balance of ecosystems. Because of their water-absorbing abilities, they help to prevent soil from drying out. In areas around the world where the soil has few nutrients, they can store and release nitrogen and phosphorous compounds many trees need and therefore increase the nutrient quality of the soil. When they grow on trees, they usually do not harm them. Yet, if they are growing on rocks, they help break down rock materials physically and chemically. Hyphae can grow into cracks and break the rock, while the acids they produce can dissolve the rock. These processes increase soil health for other creatures.

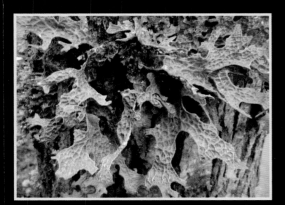

Lobaria pulmonaria or lung lichen has a green algal partner that shows through the fungus when wet (left). This is what it looks like when it is dry. It will be brittle and easily crushed (right).

BUILDING MEMORIES

For family devotions, study 1 Corinthians 12–14 and encourage one another to identify the gifts God has given you. Practice using these gifts in ways that will strengthen, protect, and encourage others. Collecting lichens as a family would also give you a wonderful opportunity to enjoy God's creation together as you focus on the kinds of relationships He has made. Here are a few collecting tips:

1. Materials needed could include a 10x hand lens, a collecting bag, and a small knife for carefully scraping lichens off of dead branches.
2. Since lichens grow slowly, take small samples.
3. Take a class or purchase a book on how to identify lichens.
4. Once collected, try to dry them as soon as you can.
5. Get permission if you are walking on someone else's property, and review all regulations about collecting creatures before hiking on local, state, or national lands.

ECOLOGIST
from the pages of history

Simon Schwendener

Simon Schwendener (1829–1919) was a Swiss plant scientist who is best remembered for his investigations about plant structure and how plants work. In 1867, he was the first to announce that a lichen was not a plant but an organism made of colonies of two different creatures, a fungus and an alga. His breakthrough was known as the Dual Hypothesis and was rejected and made fun of by many of the top lichen researchers of the day. As often happens in science, new ideas take time to be accepted. Continued research has proven Schwendener correct, with a few exceptions. Since Schwendener's time, scientists have differed about how these creatures relate to one another. Rather than a commensal or parasitic relationship, recent data suggests a mutualistic one.

Ecology in Practice: You and Ecological Research

The United States Forest Service has some educational activities you can do at this site: www.fs.fed.us/kids

Try the following Project Questions after you come back from collecting, observing, and maybe photographing lichens. From your observations, try to design an experiment to answer the following questions.

1. How fast do lichens grow?
2. What do lichens grow on?
3. How do lichens change over time?
4. What happens when two lichens meet?
5. How do lichens change from a rural area moving toward and into an urban area?

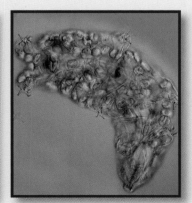

How to Find a Water Bear (Tardigrade)

1. Do an Internet or library search and read about them, making sure you know what they look like.
2. Take a small lichen sample and put it upside down in a cup of distilled water for several hours.
3. From the debris that settled on the bottom of the cup, take an eyedropper (pipet) and place a drop of water (including a little debris) on a microscope slide, and use a microscope with at least 40x magnification.
4. Explore. You might be surprised at what kinds of creatures lurk in that drop of water. For further information on designing and doing science experiments, see Appendix A.

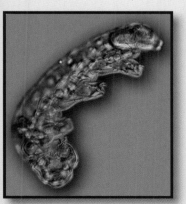
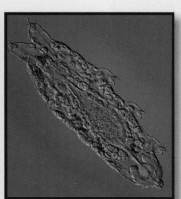

Sources

Armitage, M.A. and G. Howe. 2007. The ultrastructure of lichen cells supports creation, not evolution: a photo essay and literature review—part 1. Creation Research Society Quarterly 44 (Summer): 40–53.

Armitage, M.A. and G. Howe. 2007. The ultrastructure of lichen cells supports creation, not evolution: a photo essay and literature review—part 2. Creation Research Society Quarterly 44 (Fall): 107–118.

Douglas, A.E. 2010. The Symbiotic Habit. Princeton University Press, Princeton, New Jersey, 202pp.

http://lichen.com/biology.html, accessed 8 July 2011

Purvis, W. 2000. Lichens. Smithsonian Institute Press, Washington, D.C., pp 6–7.

In union there is strength.
— Aesop

Chapter 6

> Level 1
> Level 2
> Level 3

A Fungus Among Us

Here is a wild and wonderful fact. Did you know plants can "talk" to other creatures? Yes, it is true! Remember this weird truth the next time you walk in a forest. You see, trees and other plants are in very important relationships with fungi, and they actually talk to one another. No, you cannot hear them because they are communicating underground by chemicals, but they are still talking to each other for lots of reasons. This relationship between plants and fungi is called a mycorrhiza, pronounced my-ko-RY-za. This strange word means "fungus roots." They can be found all over the world, there are many different types, and they were made by God to do important jobs in the environment.

THE BIG FIVE

WHO	Who was Albert Bernard Frank?
WHAT	What is a mycorrhiza?
WHERE	Where are mycorrhizae found?
WHY	Why are mycorrhizae important to study?
HOW	How do mycorrhizae work together?

Words to know

Arbuscular mycorrhizae	Mycoheterotroph
Bioremediation	Orchid
Botanist	Pelotons
Ectomycorrhiza (plural ectomycorrhizae)	Phloem
Endomycorrhiza (plural endomycorrhizae)	Phytoremediation
Fungus (plural fungi)	Succession
Hypha (plural hyphae)	Symbiosis
Mycorrhiza (plural mycorrhizae)	Vascular plants
	Vascular tissue
	Xylem

Why do plants talk to fungi?

Plant communication with fungi is still being studied, but we think they chemically communicate to find one another and coordinate their body systems so they can work together.

ECOLOGIST
from the pages of history

Albert Bernard Frank

Albert Bernard Frank (1839–1900) was a German botanist (plant scientist) who first described mycorrhizae as a mutual relationship between plants and fungi. When scientists first heard his descriptions, they doubted his conclusions. They thought that there was no way that a fungus could live in a plant root without causing it harm. Their closed-mindedness slowed our understanding of the fungus-plant relationship. Nevertheless, by the late 1800s, Frank's determination showed that the fungus/plant relationship was mutual.

Varying Relational Types

Ectomycorrhiza

There are several types of mycorrhizal relationships around the world, and most relationships are considered symbiotic, with just a few that may be weakly parasitic. Generally both creatures benefit because the fungus is able to use sugars produced by the plant and the plant is able to get water and minerals that the fungus gets from the soil.

One category of mycorrhiza is an ectomycorrhiza. The prefix *ecto* means "outside" and refers to the fungus growing on the outside of the root and not the inside of root cells. The fungus grows around the outside of the root on trees like pine, oak, and birch, forming a covering or sheath with its hyphae. Occasionally, a fungus may penetrate inside a root cell. There are three fungi groups that are involved with this symbiosis; Basidiomycete (like mushrooms), Ascomycetes (like cup fungi), and Zygomycetes (like molds).

Basidiomycete

Ascomycetes

Zygomycetes

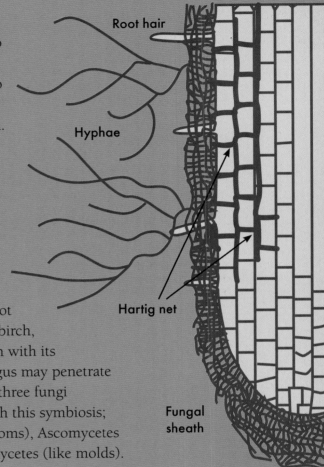
Root hair

Hyphae

Hartig net

Fungal sheath

Faucet Facts

According to some estimates, 80 to 100 gallons of water are used each day per person. And in the past few decades a lot of faucets, toilets, and showers were designed and manufactured to be more water-efficient by restricting the amount of water that flows through them. Some typical water usage includes:

• Brushing teeth – less than a gallon if the water is off while brushing

• Toilets – 1.5 gallons to 4 gallons

• Washing hands or face – 1 gallon

• Washing clothes – 25 to 40 gallons per load

You can conserve water by not leaving the water running and by doing full laundry loads instead of partial ones, or be sure to use the size options per load to help the washer be more efficient. (Water usage facts are from www.usgs.gov)

Endomycorrhiza

A second type of mycorrhiza is called an endomychorrhiza. Endo refers to "inside" and means that the fungus grows inside the cell. There are many types of endomycorrhizae and we will discuss two: arbuscular and orchid.

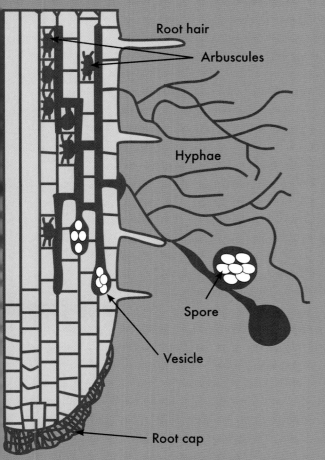

Root hair

Arbuscules

Hyphae

Spore

Vesicle

Root cap

The most common endomycorrhizae is called arbuscular. Arbuscular mycorrhizae (AM) are fungal symbionts that can only live if they are in a mutual relationship with plants. It is estimated that 80 to 90 percent of the vascular plants on the planet have endomycorrhizae. They can also be in symbiosis with some nonvascular plants, like mosses. Vascular plants are plants that have tissues that can transport water and nutrients. Xylem tissue transports water and minerals up the plant, and phloem tissue transports sugars and other materials throughout the plant. You might think of vascular tissue like the plumbing inside your house or apartment.

The word arbuscular has to do with how the hyphae form arbuscles (little branch-like tree structures) inside the cell, between the cell wall and cell membrane. The hyphae can also produce vesicles (balloon-like structures) inside the root cells. These structures help the nutrients go back and forth between the fungus and the plant. For example, sugars made by the plant can be used by the fungus, and nitrogen obtained by the fungus can be used by the plant. The type of fungi in these symbioses come from the phylum (group) called Glomeromycota and are completely dependent on the plants for their survival.

Chemical Communication

Plants chemically communicate with fungi to begin their relationship. For example, the fungus *Glomus mossae* produces hyphae that respond to certain chemicals made by plant roots by growing toward them. There is evidence that non-host plants do not produce these chemical signals, and therefore, fungal hyphae are not encouraged to form symbioses with them.

Other evidence for genuine communication and coordination between plants and fungi are seen where the root and hypha come together. Once the hypha makes contact, a flattened hyphal pressing organ punches through the outer root layer. Once inside, hyphae grow and may breech the first part of the cell wall but do not go past the cell membrane. Instead, the plant cell membrane grows around the hypha and makes a little compartment. This compartment keeps the fungus cytoplasm from mixing with plant cytoplasm. It is thought that many genes must be involved with both organisms, and what the fungus lacks can be provided by plant genes enabling them to work together. For this cooperation to happen, there must

Symbiotes and Parasites

It is estimated that about 200 plant genera may host thousands of species of ectomycorrhizal fungi, and one tree may have 15 or more different ectomycorrhizal relationships. When the ectomycorrhizal fungi form a covering or a sheath around the roots, the hyphae then form a net around the cortex cells. The cortex lies just under the outer cell layer of the root and is involved with nutrient transport and possibly food storage. These hyphae help the nutrient exchange between the plant and fungus.

What is even more fascinating is that the fungal hyphae grow extensive mats of mycelia (singular mycelium) throughout the forest soil, and these connect to root systems of all kinds of plants growing in the forest. This chemical communication between fungus and plant is very complex, and nutrients are shared between tree species. For example, it was shown that carbon nutrients moved from paper birch trees to Douglas fir trees, suggesting that ectomycorrhizae are not only bridging the gap of nutrient exchange between different tree species but also participating in forest succession. Succession refers to how ecological communities of a particular area change over time.

Indian Pipe Plants

Another interesting relationship with ectomycorrhizae involves the Indian pipe plant, *Monotropa uniflora*. The Indian pipe is a mycoheterotroph, which is a plant that does not photosynthesize but gets its carbon nutrients from mycorrhizae that get their carbon from photosynthesizing plants. This relationship is usually considered parasitic on both the fungus and plants they tap nutrients from.

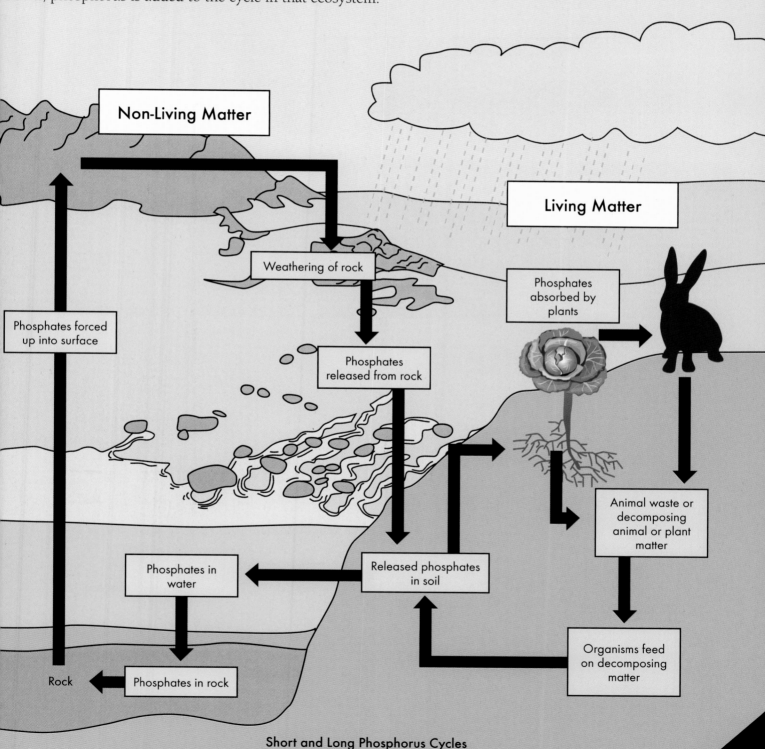

WONDER WHY

Why is phosphorus important?

Without phosphorus, life could not exist. It is an important element needed for all organism development and growth. There are two cycles — the short and the long. In the short cycle, soil phosphorus can be in the chemical form called phosphate. Phosphate is absorbed and used by producers (with the help of mycorrhizae) and are in turn absorbed and used by consumers. When creatures release wastes or die, the decomposing fungi and bacteria recycle them back into the soil. In the long cycle, phosphorus may be an ingredient in some rocks. As these rocks physically and chemically break down, phosphorus is added to the cycle in that ecosystem.

Non-Living Matter

Living Matter

Weathering of rock

Phosphates absorbed by plants

Phosphates forced up into surface

Phosphates released from rock

Animal waste or decomposing animal or plant matter

Phosphates in water

Released phosphates in soil

Rock

Phosphates in rock

Organisms feed on decomposing matter

Short and Long Phosphorus Cycles

47

Flowering Bedroom Slippers

If you know where and when to look, there might be some beautiful flowers growing near you. Some of these flowers look just like puffy bedroom slippers. These are called lady slippers, and they are in the orchid family. They come in all kinds of colors, like green, red, purple, white, pink, and colorful mixes. Did you know that we would never enjoy these beautiful creations were it not for their fungus friends? Yup, orchids would not exist if they did not have a special type of mycorrhizal relationship.

Examining the Fungus–Orchid Relationship

The fungal hyphae form tight coils called pelotons inside the orchid cells. It is where the peleton and orchid cell meet that carbohydrate absorption and movement probably take place. Cells in the orchid do not show signs of stress when they change in response to the presence of hyphae. It is unclear what is happening inside the cell, but it may be that their genetic systems are cooperating with each other so this amazing relationship can take place. There is growing evidence that orchid mycorrhizae (as with all mycorrhizae) are important in providing mineral nutrition such as nitrogen (N) and phosphorous (P) to plants.

The Rare Wonders of Vanilla

There is only one orchid variety that produces an edible pod. Orchids of the genus *Vanilla* were first cultivated in Mexico by the Totonaco people, prior to the Aztec civilization. These rare plants could only be pollinated by the Melipona bee, found only in the Mexican region. The pods, also called vanilla beans, produce flavorings used in baking, as well as a fragrance, used in aromatherapy and perfumes. It is now most often pollinated by hand, and is also produced in the Bourbon Islands, India, Indonesia, Papua New Guinea, Tahiti, and Uganda.

Orchids and Mutual Benefits

Another type of endomycorrhizae happens with orchids. As with the ectomycorrhiza, a large number of orchid mycorrhizae are Basidiomycetes and are common in soils globally. Orchid seeds are very tiny and do not have the stored food needed to germinate and grow, like most seeds have. So, in order to survive and germinate, they need a complex system of mycorrhizal symbionts to give them the nutrients they need for making it to adulthood. There is evidence that at least some orchid species are specific toward particular fungi. Once the seeds leave the parent plant, they absorb water and swell up. Fungal hyphae enter the seed coat, and the relationship may begin, but not always. Sometime the fungus may either kill the seed or nothing happens at all. If the relationship begins, the fungus can obtain important carbon nutrients from the soil and pass them on to the orchid. These nutrients are important so the orchid can both grow and carry on important life functions. The fungus may also be able to make certain vitamins and growth compounds important for the embryo. Presently, it is not known how the fungus benefits, so this may be a commensal relationship.

Stages of Growth

More than one fungal species can form pelotons in an individual orchid. Some orchids experience different fungi at different developmental stages of their growth. In the case of one orchid, *Gastrodia elata*, one fungus is needed for germination, but full orchid development can only occur if a different fungus establishes a relationship with it. Other research has found that orchids growing on the top of other plants contain both interlaced fungal hyphae and nitrogen-fixing cyanobacteria that form a sheath on the roots, making a three-way relationship.

Important Jobs God Has Given Mycorrhizae to Do

Now that you have seen that plants and fungi have very important relationships, you may wonder if God has given them important jobs to do. He sure has! These creatures help prevent their plant "friends" from getting sick, and when it has not rained in a long time, they also help their plants live through the dry times. They are also good at making poor soils healthy and helping plants grow in places that have bad chemicals that kill other plants.

The Environment and Bioremediation

Arbuscular mycorrhizae (AM) have great promise in cleaning up and improving unhealthy ecosystems and helping us to improve our agriculture. Bioremediation is the technology that uses organisms to clean polluted environments. Phytoremediation is a form of bioremediation where plants are used to take toxins out of soil, and AM have great promise in this area. Because the ability to absorb heavy metals, like zinc and cadmium, depends on fungal species, soil microbes, and plant/fungus relationships, interest in using AM to remove toxins out of polluted soils is growing. Bioremediation is also being done on ruined soils and land that has changed into desert.

AM may also help us with improving our agricultural practices. Improper farming can cause harm to ecosystems and healthy crops by destroying soil structure and overloading them with chemicals such as fertilizer and pesticides. AM forms direct links to crop plants and affects soil structure by binding soil clumps so that soil is more stable and nutrients do not get eroded away. This alone has the indirect result of growing healthier crops. May our understanding of these complex relationships help us to be better stewards of Christ's world!

Phytoremediation: The lack of vegetation in the barren area is a result of the soil's high zinc content and low pH. Alpine pennycress (inset left) doesn't just thrive on soils contaminated with zinc and cadmium—it cleans them up by removing the excess metals. Hyperaccumulators like *Thiaspi* (inset right) possess genes that regulate the amount of metals taken up from the soil by roots and deposited at other locations within the plant.

BUILDING MEMORIES

As a family devotion, read Genesis 1 and 2, and focus on the creation of plants and some of the revealed purposes for them. Over a few days, assign different family members to do a Scripture search for plants/trees, and let them share how God uses them as spiritual analogies for our Christian walk.

Then research the plants in your area that are known to have mycorrhizal relationships, and with a plant field guide, go on a family "treasure hunt" for these plants. If they are not endangered or a local species of concern, take a leaf, stem, root, fruit, and/or flower sample and preserve them.

Collecting and Preserving Mycorrhizal Plants as a Family — Using pruners or scissors, cut a leaf, stem, root, fruit, and/or flower sample(s) from common plants known to have mycorrhizae. On the newspaper, carefully lay the samples out the way you want them. Take the newspaper and carefully fold it over the samples in the positions you want them, and place the folded newspaper, with samples, inside a book. Place several books or other weights on top of the book containing the newspaper with samples, and let them sit for two or three weeks. After a few weeks, open the book and carefully unfold the newspaper with samples, and you should have well-preserved plant specimens. You can now carefully mount the plant specimens on white construction paper (you can also get plant-mounting paper through science outlets like Wards or Carolina Biological Supply), and in the corner of the white paper, record the following information: collector, date/time/location of plant when collected, common name of plant, family name of plant, scientific name of plant (genus and species), and type of mycorrhizae. These mounted plants can now be hung on your walls, and they should last a long time.

Older students can take this project further and combine the plant specimens with a major report about the plant and the type of mycorrhizae it is associated with. Added information would be how the mycorrhizal relationship helps the plant in the ecosystem, whether it is helping other plants at the same time, and thoughts about how we can better steward the ecosystems the plant grows in.

MAKE A DIFFERENCE • ECO-FRIENDLY

Water Wars

A drought is an extended period of time (months to even years) when an area has a deficient water supply, from lack of rainfall or depletion of surface or groundwater sources. This negatively impacts crops, raising livestock, and even can put a stop to filling swimming pools.

Another challenge is water systems that are used in large cities, though the water is transported from distant sources, meaning water from these areas is not available to people who may live closer than the city itself. This conflict has led to "water wars" in the past, and presents a difficult challenge for the future.

"We get closer to God as we get more intimately and understandingly acquainted with the things He has created."
— George Washington Carver

Chapter 7

> Level 1
> Level 2
> Level 3

The Plant — Bacteria Connection

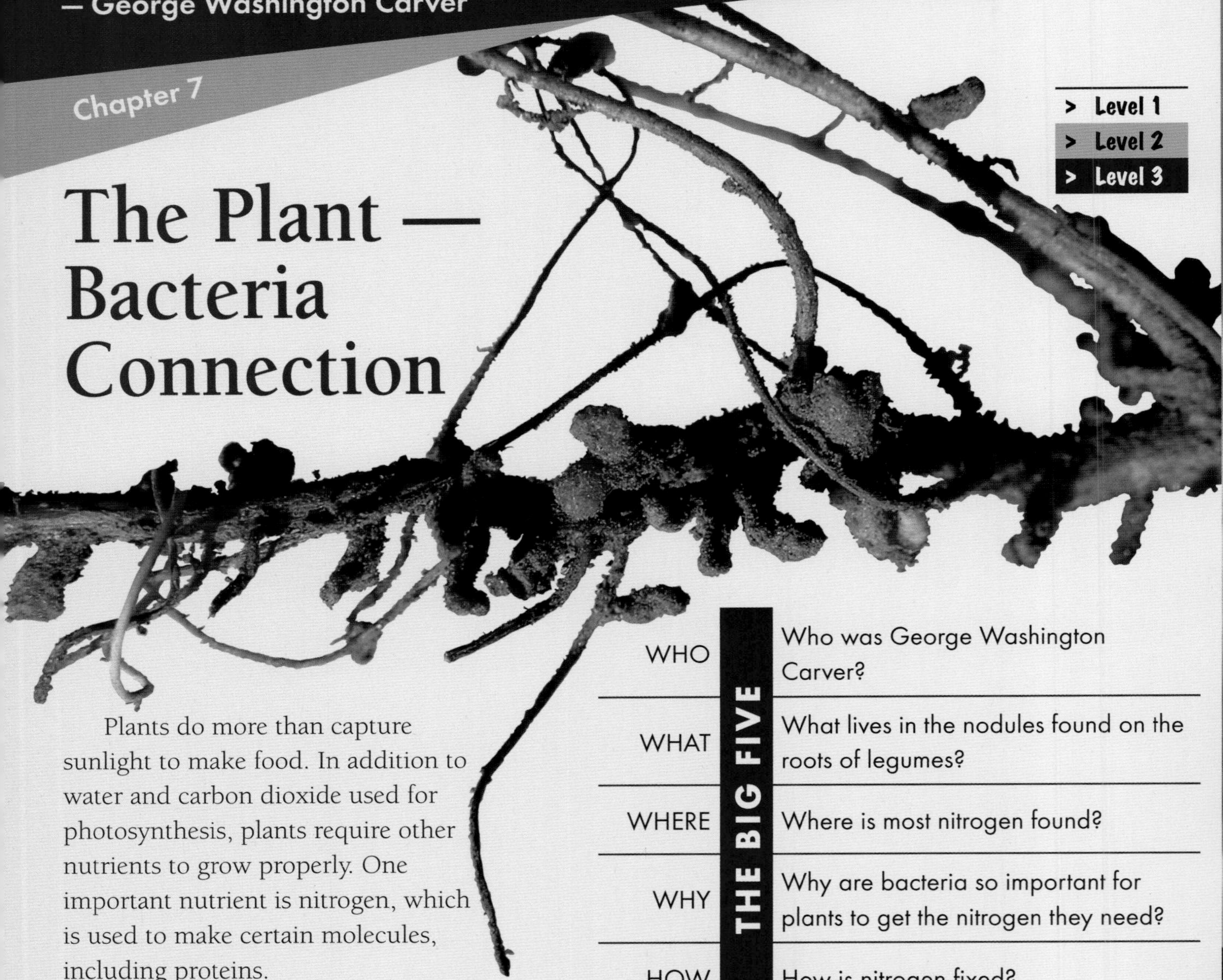

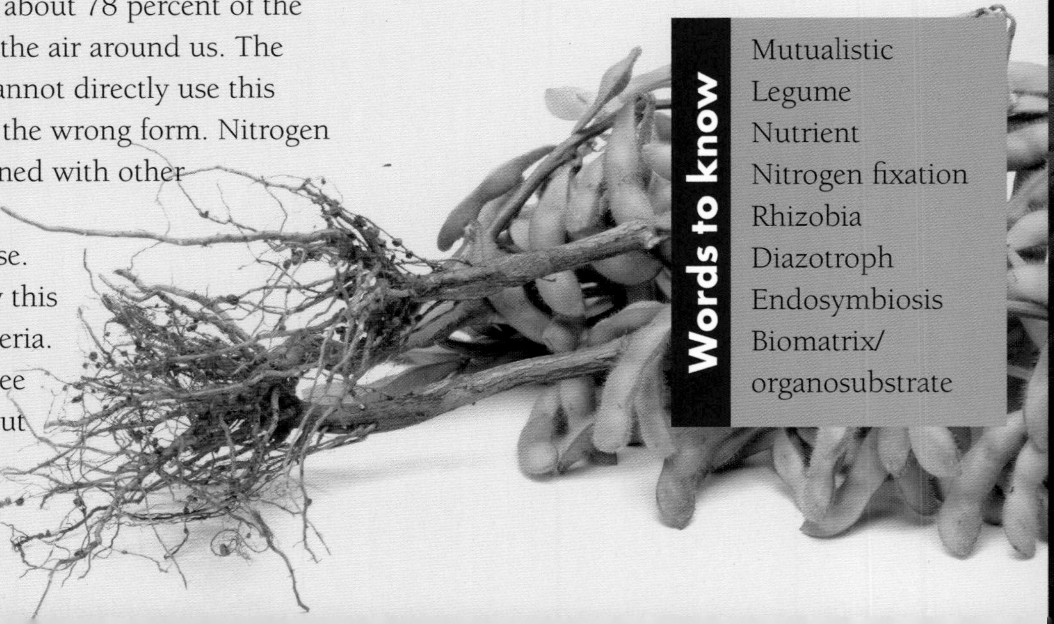

THE BIG FIVE	
WHO	Who was George Washington Carver?
WHAT	What lives in the nodules found on the roots of legumes?
WHERE	Where is most nitrogen found?
WHY	Why are bacteria so important for plants to get the nitrogen they need?
HOW	How is nitrogen fixed?

Plants do more than capture sunlight to make food. In addition to water and carbon dioxide used for photosynthesis, plants require other nutrients to grow properly. One important nutrient is nitrogen, which is used to make certain molecules, including proteins.

Nitrogen occurs as a gas in air. It is abundant, making up about 78 percent of the total volume of gases in the air around us. The problem is that plants cannot directly use this nitrogen because it is in the wrong form. Nitrogen must be fixed, or combined with other elements, for it to be in a form that plants can use. One very important way this happens is through bacteria. Bacteria are too tiny to see without a microscope, but they are essential for life on earth.

Words to know

Mutualistic
Legume
Nutrient
Nitrogen fixation
Rhizobia
Diazotroph
Endosymbiosis
Biomatrix/organosubstrate

How Can You Fix Nitrogen?

Nitrogen gas (N_2) in the atmosphere consists of nitrogen atoms paired together by very strong triple bonds. This makes the gas inert (unlikely to react with other molecules) and unusable for plants and animals. There are two main ways nitrogen can be fixed. First, nitrogen can be combined with hydrogen to form ammonia (NH_3) or related ions. There are a variety of bacteria that can accomplish this. These bacteria are referred to as diazotrophs (*azo* comes from the French word for nitrogen, *azote*). They have special enzyme systems that allow them to directly use nitrogen gas, and in doing so, they make it available to plants as well.

A second way to fix nitrogen is to combine it with oxygen. Lightning releases tremendous energy, splitting nitrogen gas (N_2) and oxygen gas (O_2) to form nitric oxide (NO). This can combine with more oxygen to form nitrogen dioxide (NO_2), a gas that readily dissolves in water. This dissolved form can fall to earth in rain. From there, bacteria known as nitrifying bacteria can change it into ammonia, which is usable by plants.

WONDER WHY *Why does nitrogen need to be fixed?*

Fixing nitrogen refers to changing nitrogen gas (N_2) into a more usable form by combining it with other elements, such as hydrogen or oxygen. It first requires that the very strong triple bond that holds the atoms together be broken. Plants and animals need nitrogen, but they cannot obtain it directly from nitrogen gas. Bacteria are important in fixing nitrogen so plants can use it. Then animals can get the nitrogen they need from plants.

Nitrogen-Fixing Bacteria

Bacteria that fix nitrogen live in the soil. Many just live near the roots of plants and help supply nitrogen. There are certain nitrogen-fixing bacteria — rhizobia — that form a very close relationship with plants known as legumes. Legumes include bean plants, pea plants, clover, and alfalfa. These plants will form small nodules (little balls) on their roots, where rhizobia live and fix nitrogen for the plant to use. Through this mutualistic relationship, the legume plant is provided with usable nitrogen. However, the benefits do not stop there. When legumes are grown, the soil is enriched in usable forms of nitrogen that can benefit other plants too. When the rhizobia work symbiotically with legumes, the amount of fixed nitrogen added to the soil is usually much greater than when other bacteria fix nitrogen separately from plants.

The Planetory Nebula NGC 2818. The red represents nitrogen coming from the clouds.

Non-Legume Bacteria Symbiosis

There are a few species of bacteria that can form a close relationship with plants that are not legumes. For example, *Frankia*, a genus of filamentous, nitrogen-fixing bacteria, are found in the nodules of the alder tree. While the details of these relationships differ from those of legumes, the overall result is quite similar: nitrogen is made more readily available to the plant.

A whole alder tree root nodule.

A sectioned alder tree root nodule.

Nitrogen Cycle

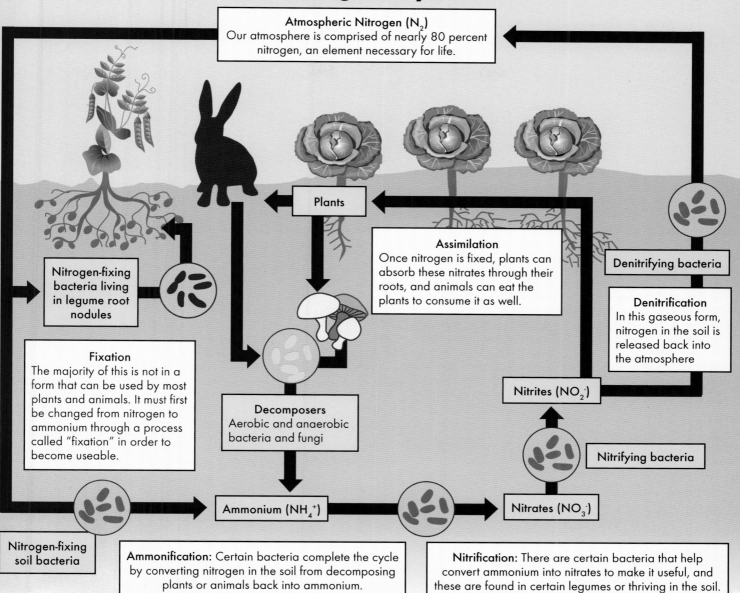

Atmospheric Nitrogen (N$_2$)
Our atmosphere is comprised of nearly 80 percent nitrogen, an element necessary for life.

Plants

Nitrogen-fixing bacteria living in legume root nodules

Assimilation
Once nitrogen is fixed, plants can absorb these nitrates through their roots, and animals can eat the plants to consume it as well.

Denitrifying bacteria

Denitrification
In this gaseous form, nitrogen in the soil is released back into the atmosphere

Fixation
The majority of this is not in a form that can be used by most plants and animals. It must first be changed from nitrogen to ammonium through a process called "fixation" in order to become useable.

Decomposers
Aerobic and anaerobic bacteria and fungi

Nitrites (NO$_2^-$)

Nitrifying bacteria

Nitrogen-fixing soil bacteria

Ammonium (NH$_4^+$)

Nitrates (NO$_3^-$)

Ammonification: Certain bacteria complete the cycle by converting nitrogen in the soil from decomposing plants or animals back into ammonium.

Nitrification: There are certain bacteria that help convert ammonium into nitrates to make it useful, and these are found in certain legumes or thriving in the soil.

Reflections

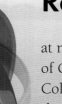

The nitrogen cycle is one of many natural cycles. When you look at natural cycles, do you see evidence of a Creator? If so, what kind of Creator? How do your thoughts compare with Jeremiah 14:22 and Colossians 1:16–17? How can understanding the relationship between rhizobia and legumes be helpful to Christians exercising biblical stewardship of creation?

Relationships Begin with Dialogue

The mutualistic relationship between legumes and rhizobia is called endosymbiosis because the bacteria actually live inside the legume roots. This close relationship does not happen by accident. Instead, a chemical dialogue is necessary for the relationship to be established.

Legume roots exude flavonoids. These are sensed by rhizobia in the surrounding soil, initiating the production of signaling molecules known as nodulation (nod) factors. These nod factors are detected by the plant roots. The plant root hairs curl around the bacteria. The bacteria enter the plant through an infection thread where they are guided into the forming nodule. The infection thread is provided by the plant and has a tunnel through which the bacteria travel to gain access.

The signaling continues. The bacteria are transformed within the nodule to bacteroids. They remain inside the plant cell surrounded by a thin membrane. The plant controls the level of oxygen and other factors so the bacteroids can fix nitrogen. While some bacteria can fix nitrogen apart from plants, rhizobia require a symbiotic relationship with plants to do so.

Some of the details vary depending on the species involved, but the successful formation of nodules containing nitrogen-fixing rhizobia (NFR) involves very complex signaling. Dozens of genes have already been identified that play an important role in the signaling and nodule formation; there are certainly many others waiting to be discovered.

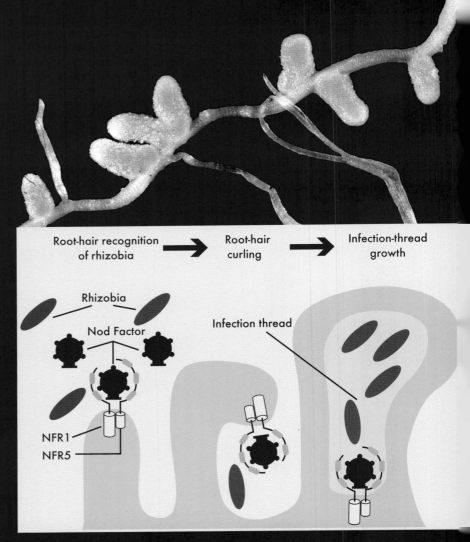

Root-hair recognition of rhizobia → Root-hair curling → Infection-thread growth

Rhizobia

Nod Factor

Infection thread

NFR1

NFR5

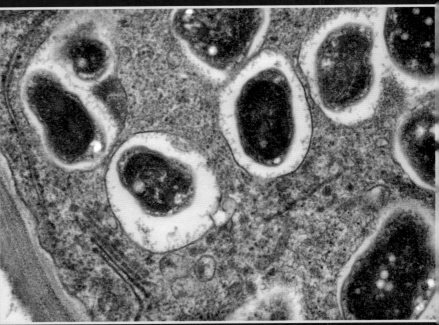

Cross section through a soybean (*Glycine max 'Essex'*) root nodule. The bacterium, *Bradyrhizobium japonicum*, colonizes the roots and establishes a nitrogen-fixing symbiosis. This high magnification image shows part of a cell with single bacteroids within their symbiosomes. In this image, endoplasmic reticulum, dictysome, and cell wall can be seen.

- Animal (cow/horse) manure
- Cardboard rolls
- Clean paper
- Coffee grounds and filters
- Cotton rags
- Dryer/vacuum cleaner lint
- Eggshells
- Fireplace ashes
- Fruits and vegetables
- Grass clippings
- Hair and fur
- Hay and straw
- Houseplants
- Leaves
- Nut shells
- Sawdust
- Shredded newspaper
- Tea bags
- Wood chips
- Wool rags
- Yard trimmings

Things Not to Compost

- Black walnut tree leaves or twigs
- Coal or charcoal ash
- Dairy products and eggs
- Diseased or insect-ridden plants
- Fats, grease, lard, or oils
- Meat or fish bones and scraps
- Pet wastes (e.g., dog or cat feces, soiled cat litter)
- Yard trimmings treated with chemical pesticides

Getting Nitrogen to the Plant

Bacteria play an important role in providing plants with usable nitrogen through nitrogen fixation. However, there is another way bacteria are important in providing nitrogen: through decomposition. Plants and animals use nitrogen in amino acids (which make up proteins) and nucleic acids (DNA and RNA). Certain bacteria break these molecules apart so the nutrients can be recycled. This cleans up waste and dead remains while releasing nutrients for life to continue.

Compost, often made from manure, is one method for adding nitrogen and other nutrients to soil used in agriculture. As the compost is broken down by bacteria and other microbes, nutrients are released that benefit plants. Another means for adding nutrients is to plant a second crop after the first is harvested. This second crop covers the ground to prevent erosion, which is why it is commonly called a cover crop. Before the main crop is planted the next year, the cover crop is plowed back into the soil so the nutrients get recycled. This kind of crop is also called a green manure crop, since green plant material is being used like manure is sometimes used. Legumes make excellent cover crops since they primarily use nitrogen from the air, with the help of rhizobia. This nitrogen is then added to the soil when the crop is plowed under.

You might wonder why, since nitrogen is essential for life, so much of it is tied up in a form that plants cannot directly use. Why not just have an abundant source of nitrogen in the form of ammonia, which plants can use? The answer is because ammonia in high doses is toxic to life. In fact, all forms of fixed nitrogen can easily become toxic if their levels get too high. Nitrogen gas is a very safe form that acts as a reservoir. Bacteria provide a necessary bridge between the nonliving environment and life. This bridge between the living and nonliving has been called an organosubtstrate or biomatrix.

WONDER WHY

Why do some farmers and gardeners rotate crops?

It was noticed that when the same plants are grown year after year on the same plot of ground, the plants tend to become less productive. The harvests in later years got noticeably smaller because the plants were using the nutrients in the soil faster than they were being replaced. Since various plants can differ somewhat in their nutrient requirements, changing what crop is planted can be helpful in allowing the soil time to recover. If a legume is planted, the soil will be enriched with nitrogen in a form that plants can use. This is because legumes use nitrogen mostly from the air. After the crop is harvested, the remainder of the plant will decompose and add extra nitrogen to the soil. This is why it is very common for a legume to be included in crop rotation.

WONDER WHY

Why did God tell the Israelites not to plant crops every seventh year?

This command (Ex. 23:10, 11; Lev. 25:2–7) is similar to the command that God's people observe a weekly Sabbath (Ex. 20:11; 23:12). There are probably several reasons:

- God wants His people to trust Him. It is essential to work and plant crops to have food. However, it is also important to take time to honor God. In obeying God, His people would learn that God is ultimately our provider and we can trust that He will provide for our needs when we obey (Matt. 6:33).

- It was a testimony to unbelievers. It would be very noticeable to others that they were behaving differently. Some people would mock them. Others would want to ask questions. In answering these questions, God's people would be pointing others to God.

- It provided the land with a rest. Just as the weekly Sabbath gave people a chance to rest and be refreshed (Ex. 23:12), the Sabbath rest for the land would have a similar effect. The yearly removal of a harvest from the field slowly depletes the soil because the harvested crop has nutrients that are not returned to the land. The Sabbath for the land gave it a chance to recover. It was a good practice to ensure the long-term productivity of the land.

BUILDING MEMORIES

Make a family garden and plant bean seeds and/or other legumes. After harvest, join together and give thanks to our God as you enjoy His wonderful bounty. You can also pull up the plant roots and check out the bacterial nodules. Be awed at the complexity of this system and how it is crucial to many ecosystems of the world.

Make Your Garden Grow

Developing a family garden can be a wonderful way to transform your own yard. You can also help produce a wonderful variety of delicious vegetables and fruits for your family and friends to enjoy. And if you rotate your plants it can help keep your garden healthy and thriving. Rotating crops can benefit your garden by

1. promoting healthy plants by replenishing the soil nutrients they need
2. helping guard against various plant diseases that can reside in the soil
3. protecting plants against certain plant pests that return each year.

The following three-year rotation cycle can be utilized for most gardens. Your garden would be separated into three sections: A, B, and C. All plants in the first plot could be planted in section B the next year and section C the last. Every fourth year you would start from the beginning. This simple plan will help your garden flourish for years to come. Remember to share your abundance!

Section A: beets, carrots, celery, cucumbers, fennel, garlic, leeks, melons, parsnips, potatoes, pumpkins, squash, tomatoes, and zucchini

Section C: broccoli, Brussels sprouts, cabbages, cauliflower, kale, kohlrabi, mustard, radishes, and turnips

Section B: artichokes, beans (bush, Lima, pole, soy), lettuce, okra, peanuts, peas, spinach, sweet corn, and Swiss chard

ECOLOGIST
from the pages of history

George Washington Carver

George Washington Carver (1864 – 1943) was a scientist who understood that legumes restore nutrients to the soil. Legumes are higher in protein than other plants, and so the peanut, a legume easy to grow in the South, helped restore fertility to the soil and provided an inexpensive source of protein for hungry people.

Born a slave, he worked his way through college in the North and then returned to the South, desiring to devote his life to improving the quality of southern farm lands and the economic prosperity of his people. As a faculty member at the Tuskegee Institute in Alabama, he turned down a number of much more lucrative offers, as the fame of his genius as an agricultural chemist spread. He developed over 300 products from the peanut and over 118 from the sweet potato.

Carver was also a sincere and humble Christian, never hesitating to confess his faith in the God of the Bible and attributing all his success and ability to God. In 1939 he was awarded the Roosevelt medal, with the following citation: "To a scientist humbly seeking the guidance of God and a liberator to men of the white race as well as the black" (from *Men of Science, Men of God*).

The Issue of Fertilizers and Stewardship

The use of fertilizers in agriculture can be controversial. On one hand, they add critical nutrients such as nitrogen, phosphorus, and potassium to the soil. This greatly increases the productivity of agricultural crops. On the other hand, nitrates and phosphates can cause serious problems for people and animals.

The nitrogen compounds in fertilizer, manure, and plant waste are broken down by microbes into nitrates. Nitrates are easily washed out into streams or leached into ground water, where they can pose a significant health risk when the water is used for drinking. Phosphates tend to stick to soil particles, but they can enter surface water through erosion. While not directly dangerous to humans, high levels promote excessive growth in algae, which can use up the oxygen supplies in the water and kill the fish.

How can we manage things so there is plenty of food for people to eat, but we avoid the problems associated with nitrates and phosphates? This is an important question, and the answers may vary according to the local climate and conditions. We need to be concerned with issues such as the long-term productivity of the land. This is sometimes known as sustainability. The long-term welfare of humans is dependent on sustainability, as is the long-term welfare of animals. As we take the time to notice the relationships around us and learn what factors affect them, we will be able to be wise stewards that use the resources God has given us for the glory of God and benefit of man.

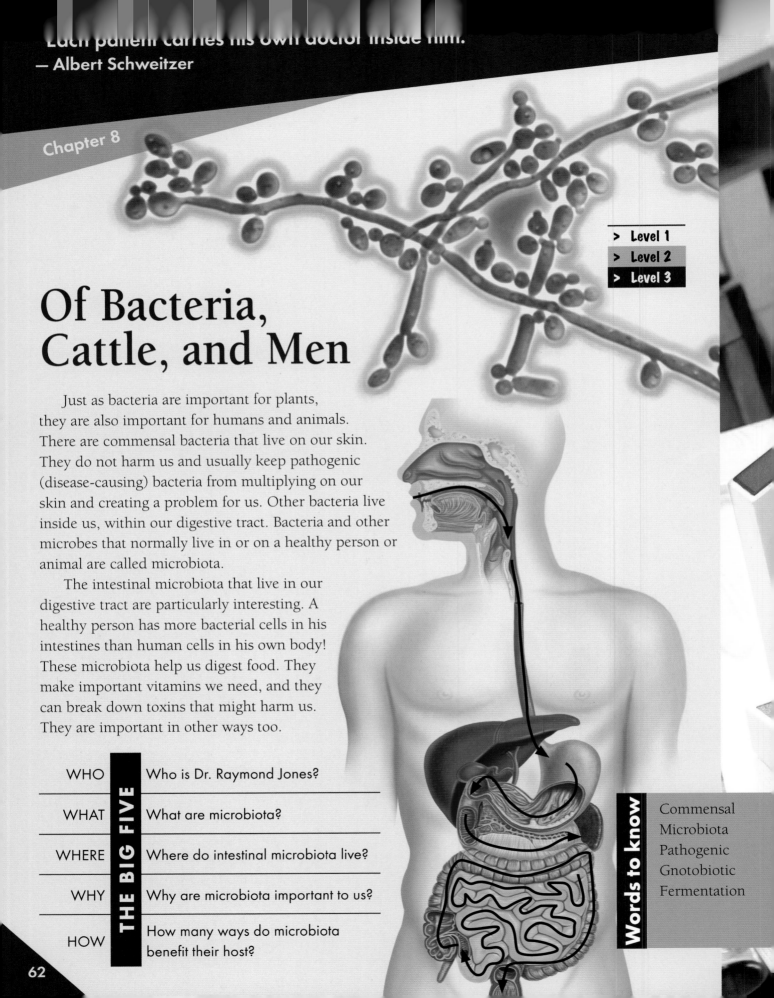

> Each patient carries his own doctor inside him.
> — Albert Schweitzer

Chapter 8

> Level 1
> Level 2
> Level 3

Of Bacteria, Cattle, and Men

Just as bacteria are important for plants, they are also important for humans and animals. There are commensal bacteria that live on our skin. They do not harm us and usually keep pathogenic (disease-causing) bacteria from multiplying on our skin and creating a problem for us. Other bacteria live inside us, within our digestive tract. Bacteria and other microbes that normally live in or on a healthy person or animal are called microbiota.

The intestinal microbiota that live in our digestive tract are particularly interesting. A healthy person has more bacterial cells in his intestines than human cells in his own body! These microbiota help us digest food. They make important vitamins we need, and they can break down toxins that might harm us. They are important in other ways too.

	THE BIG FIVE	
WHO		Who is Dr. Raymond Jones?
WHAT		What are microbiota?
WHERE		Where do intestinal microbiota live?
WHY		Why are microbiota important to us?
HOW		How many ways do microbiota benefit their host?

Words to know

Commensal
Microbiota
Pathogenic
Gnotobiotic
Fermentation

62

Why do some people recommend eating yogurt after being on antibiotics?

Yogurt is made with the help of certain bacteria. These bacteria are friendly and can help replace friendly bacteria that may have died from the antibiotics that were intended to kill the pathogenic (harmful) bacteria. As long as the yogurt has not been heat treated, which is sometimes done to extend shelf life or reduce tartness, it is assumed that the bacteria will remain alive. These yogurts may carry the "Live and Active Cultures" seal.

Unfortunately, this doesn't always guarantee they are still alive when they are eaten; tests run on store-bought yogurt show that the bacteria often die by the time they are available at the grocery store.

Bath or Shower?

MAKE A DIFFERENCE ECO-FRIENDLY

Decide for yourself — shower or bath? Showers seem a water-saving option, but do your math first! A full bathtub averages around 36 gallons while a shower uses 2 to 5 gallons a minute depending on the age of the showerhead. If you have an older showerhead and are in the shower for 15 minutes, you could be using around 75 gallons of water. Visit www. usgs.gov for more information about water and conservation efforts.

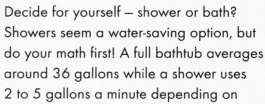

Establishing a Healthy Microbiotic Community

Exposure to bacteria and other microbes begins at birth, when a baby leaves the sterile environment within its mother. People and animals come equipped with the means to interact with these microbes through their immune systems. The bacteria "talk" with the immune system through a variety of chemical compounds. This dialogue is necessary so a healthy community of microbiota can become established and maintained.

The microbiotic community is made up of several hundred to a thousand or more different species. The specific species and their numbers can vary over time and are closely related to the diet of the host. Herbivores tend to have a much different array of species than omnivores or carnivores. The host benefits the microbiota by providing a stable environment for them to live. The microbiota benefit the host in many ways. In addition to helping with digestion, intestinal microbiota are necessary for proper development of the digestive tract. They are also important for the immune system to develop properly. Laboratory animals that were kept in sterile conditions so they were never exposed to any microbes were more sensitive to stress and had difficulty learning.

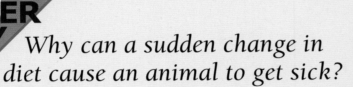

WONDER WHY *Why can a sudden change in diet cause an animal to get sick?*

To effectively digest different foods, different bacteria are important. While the microbiotic community can change over time, a sudden, dramatic change in diet can disrupt the healthy balance that existed. This may cause the host to become sick. This is why it is recommended that dietary changes be made more slowly, so the microbiota have time to adjust.

Microbiota and Digestion

Bacteria can break down carbohydrates for energy through a process known as fermentation. For example, certain bacteria are used to break down lactose in milk to produce yogurt. The sour taste in plain yogurt is from the lactic acid that the bacteria produce. This type of fermentation is also used to make sauerkraut and pickles.

Some members of the microbiota, mostly certain bacteria, have special enzymes that allow them to break down carbohydrates that are difficult or impossible for animals to break down. When these microbes break down these otherwise indigestible components of the diet, they produce short-chain fatty acids, which can be used as an energy source by the host.

Humans receive less than 10 percent of their total energy from short-chain fatty acids. Many herbivores receive far more. This is because they rely on microbiota that can digest cellulose and other indigestible carbohydrates in leaves, grass, and bark. Herbivores have a special design to their digestive tract to allow for a healthy population of these fermenting microbiota. There are two basic strategies that are used: foregut fermentation and hindgut fermentation.

The Missing Microbe

Dr. Raymond Jones is a Christian and creationist who worked for 38 years for CSIRO, a respected Australian government research organization. He published around 140 scientific papers as he worked in agriculture, especially the area of tropical grazing. He is probably best known for solving an interesting mystery surrounding the plant Leucaena.

Leucaena is a shrubby tree in the legume family that was introduced to Australia to help boost beef production in the seasonally dry tropics. This legume retains its nutritional content in the dry season when most other plants lose much of their feed value. A problem soon became apparent when cattle grazing Leucaena began to waste away, becoming thin. There was a toxin in the plant making these animals sick.

Dr. Jones proposed that the cattle were missing a microbe necessary to break down the toxin in this plant. He met a surprising amount of resistance to this idea, as most people had assumed that the microbiota in cattle were always the same. He was able to demonstrate that he was correct by introducing a specific microbe from other ruminants that could eat *Leucaena* with no ill effects. The bacterium was named *Synergistes jonesii* in his honor.

Up Front about Fermentation

Foregut fermentation is where one or more portions of the stomach contain a holding tank where the microbiota ferment carbohydrates, such as those found in grasses or leaves, which the animal cannot digest on its own. Ruminants are best known for this type of digestion. The ruminant stomach is large and divided into four chambers. As the animals graze, the grass is swallowed and goes to the largest chamber of the stomach, the rumen. Later, when the animal is resting, it can be brought back up into the mouth to be re-chewed so the microbiota will have more surface area to work on as they continue digesting it. This process is known as rumination, or more commonly called "chewing its cud."

Ruminants not only ruminate, but they also have a unique design to their stomachs. Each of the four chambers has a different distinctive lining, and only the fourth chamber has digestive glands. Essentially, the microbiota digest all the food, and then the ruminant digests the microbiota. Ruminants receive most of their energy from the short-chain fatty acids produced by the microbiota. They also benefit from the B vitamins produced by the bacteria and acquire necessary amino acids when they digest the microbes.

Camels, alpacas, and llamas also ruminate, but their multi-chambered stomachs lack some distinctive features of ruminants. For example, they have only three chambers, and the lining differs from that of ruminants. Still, they benefit from the short-chain fatty acids, B vitamins, and essential amino acids much like ruminants do.

Many other animals also use some degree of foregut fermentation, even if they do not ruminate. Sloths, peccaries, kangaroos, hippos, leaf-eating monkeys, dolphins, and whales are considered foregut fermenters. While they do not extract as much energy from the food as ruminants, microbial digestion still plays a large role in helping them obtain energy from their diet.

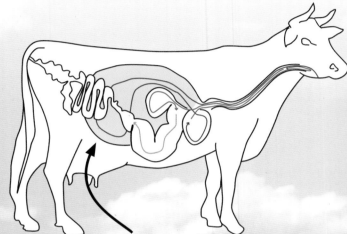

Foregut fermenters: Process occurs in the stomach as microbiota ferment carbohydrates.

Foregut fermenters

Water buffalo

New Zealand sheep

Goat

Mule deer

A Little Behind on the Process

Other animals use microbiota for fermentation in portions of the digestive tract after the stomach. These are known as hindgut fermenters. In this case, the fermentation occurs in the colon and/or cecum. The cecum, a blind pouch much like the human appendix, is well developed in many hindgut fermenters. Here the microbiota can digest tough plant material that the host cannot. In this way, the host may be able to get over 40 percent of its energy from the short-chain fatty acids produced by these microbes.

Hindgut fermentation may also take place in the colon. In this case, the colon is well developed, with many sacs where the microbiota work. Animals that use hindgut fermentation include rhinos, horses, tapirs, elephants, manatees, herbivorous apes, herbivorous rodents, koalas, rabbits, and hares.

One disadvantage of hindgut fermentation is that many of the B vitamins and amino acids are not well absorbed before they exit the digestive tract. Some animals, like the rabbit, have a way around this. When the cecum empties its contents, a soft pellet, called a cecotrope, is passed out of the animal. This nutrient-rich pellet is re-eaten, providing the animal with important B vitamins and amino acids. The animal also passes a drier fecal pellet, which is not normally re-eaten. In this way, the rabbit, much like a deer or cow, re-chews its food.

Some animals can vary their digestive systems somewhat to accommodate different food sources. For example, the ruffed grouse and willow ptarmigan can vary the length of their intestines and/or ceca (plural of cecum; this organ is paired when it occurs in birds) in response to seasonal changes in diet. A cecum is a "pouch" off of the large intestines containing microflora.

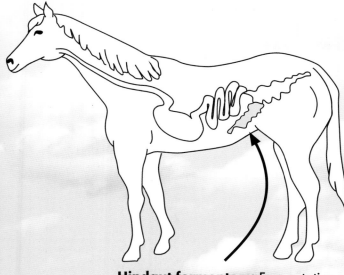

Hindgut fermenters: Fermentation occurs in the colon and/or cecum.

Hindgut fermenters

Horse Elephant Coypus Wild rabbit

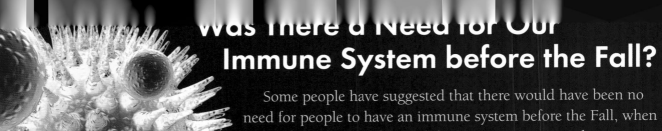

Was There a Need for Our Immune System before the Fall?

Some people have suggested that there would have been no need for people to have an immune system before the Fall, when creation was still perfect and there were no nasty pathogens. However, this assumes the only purpose of the immune system is to seek and destroy pathogens. The immune system can be better understood as a system by which an animal communicates with its environment. It plays an essential role in establishing relationships with microbiota and maintaining a healthy microbiotic community.

Understanding Microbiota

There is such a complex variety of organisms found in the microbiota that it is often difficult to determine which microbes are having which effect. To simplify studies, gnotobiotic animals are often used. Gnotobiotic, from the Greek words for known (gnostos) and life (bios), are animals that have only been exposed to microbes that the researcher has chosen to expose them to.

Normally people and animals are exposed to microbes at birth. To prevent this, gnotobiotic animals are surgically removed by cesarean section in a sterile environment. They are then kept in a special sterile environment. At this point, the young animals have not been exposed to any microbes and are known as germ-free. Later, the researcher can choose to expose them to specific microbes. The animals are still kept in an otherwise-sterile environment so no other microbes are introduced. These gnotobiotic animals have contributed much to our understanding of the importance of microbiota.

Research has shown that germ-free animals have significant differences in their digestive tract from lack of exposure to microbiota. Some cells in the digestive tract do not develop normally, and the cecum is greatly enlarged. These features are reversed when bacteria are introduced.

Some other surprising results are that germ-free animals have a more dramatic stress response. They have higher anxiety when stressed. They also have defects in learning, with or without stress.

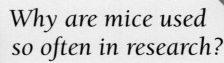

Why are mice used so often in research?

WONDER WHY

Small rodents, such as mice and rats, make excellent research animals because they are small and are quite content to live in a confined space where there is plenty of food. They are mammals and fortunately for us, have many similarities to humans and other mammals. There have been countless discoveries made using these laboratory animals that have been important in advancing both human and veterinary medicine. One of the areas where they have contributed to our knowledge is on the importance of microbiota.

Ethical Concerns of Animal Research

Genesis 1 makes it clear that God created plants and animals according to their kinds. Mankind was created to rule over the animals (Gen. 1:26, 28). Abel kept flocks (Gen. 4:2), which he used when he brought a sacrifice to God. Although eating meat was not permitted until much later (after the Flood; Gen. 9:2–3), Abel may have used milk or wool from his flocks (Deut. 32:14; 1 Cor. 9:7; Lev. 13:47–48). It is clear that keeping animals and using them for meat, milk, and wool was acceptable later. However, it was also expected that someone who kept animals would adequately care for them (Prov. 12:10). In fact, the Bible paints a very negative picture of someone who expects to benefit from his animals without even caring about their needs (Ezek. 34:2–5).

People who have not worked with particular animals may lack the ability to distinguish between humane and inhumane treatment. For example, baby pigs are often castrated when they are young. This has several advantages, including reducing fighting and the injuries associated with it. The procedure takes around 15 seconds with an experienced worker and a sharp knife. While the baby pig will squeal while being handled and certainly feels some pain, when it is set back down in the pen, it will generally go back to what it was doing, often going over to eat. Some have suggested it would be more humane to use anesthesia. However, anesthesia is far more stressful, and the baby pig would take much longer to recover.

When animals are used for research, we are using them to learn so we can be better managers of the resources God has given us. Knowledge acquired from this type of research not only benefits humans but other animals as well. Is it acceptable to God to use animals in research in this way? In a biblical worldview, there is strong reason to answer yes. God Himself set up the sacrificial system that was used for well over a millennium in Israel. By faith, devout Jews sacrificed animals as commanded. It served to teach the people that sin separates us from God, and we are in need of forgiveness. It served to teach that sin brings death, and only the death of an innocent substitute can restore our relationship to God (Lev. 17:11). While the sacrifices themselves did not take away the sins of the people (Heb. 10:1–4), it enabled those who trusted in God to recognize Jesus, the Lamb of God, who did come to take away sins (John 1:29). Thus, the animals that were sacrificed served to teach the people so they might be brought into right relationship with God.

Chapter 9

To the Root of the Matter

> **Level 1**
> **Level 2**
> **Level 3**

THE BIG FIVE

WHO	Who can explore the world around them?
WHAT	What is symbiosis?
WHERE	Where can we find a foundation for everything in life?
WHY	Why is a biblical worldview necessary for science?
HOW	How do people make sense of what they see in the world around them?

The Bible provides a solid foundation for understanding everything in life. It gives us a basis for scientific investigation. In science, we use our senses to observe the world around us. When we do this, we must assume that our senses are basically reliable. We must also assume that we can understand our world. The Bible tells us that God created people to rule over and care for the earth (Gen. 1:26–28; 2:15). God created us for this purpose, so it makes sense that our senses would usually be reliable and that we can understand our planet. This way we can learn about and care for God's creation.

Scientists have a word for organisms living together. That word is symbiosis, which refers to one or more organisms in long-term relationships where each organism is benefited by the partnership of the others. It comes from two Greek words: *syn*, meaning "with," and *bios*, meaning "life."

Words to know

Symbiosis
Organisms
Pantheism
Recycled

The Fascinating World around Us

There are so many different kinds of plants and animals. There are so many people with different skills and personalities. It is interesting to see how they interact with each other. In this book we have explored much of the variety in the world around us. From the people we see each day to a variety of plants and animals to tiny organisms that can only be seen through a microscope, we explored life around us and saw how it was well designed to live together.

How will we make sense of what we see? All people have a way of looking at and understanding the world around them. The way we view the world is called a worldview. A worldview helps us give meaning to what we see in the world around us. To understand our observations, we have used a biblical worldview. This means we consider what we see in light of the history and teachings of the Bible.

Worldviews, Science, and Life

Our worldview determines how we look at and value the world around us. It stems from how we answer basic questions regarding human identity: Who am I? Where did I come from? What is my purpose? Our answers to these questions influence what we think of ourselves, God, other people, and the world around us. Our beliefs about these topics influence how we will behave.

In secular education, an atheistic worldview is usually presented in science classes. The question, "Where did I come from?" is answered with evolution. In other words, humans are considered to be just animals that resulted from chance mutations and natural selection. If there is a god, he is never allowed to play any real role in the story. Unfortunately, the history presented leaves no basis for valuing human life or having a sense of purpose. It is not that an evolutionist cannot choose to value life or have a purpose; it is just that his beliefs do not provide a logical basis for doing so. In contrast, a biblical worldview teaches that we were created by a loving God who has purpose for our lives. The fact that humans were specially created in God's image provides the basis for placing a high value on human life.

Ironically, although the foundation of science depends on concepts that come directly from a biblical worldview, "science" is often used as an excuse to reject the history and teachings in the Bible. If one looks more closely, it becomes apparent that science is not in conflict with the Bible; rather, some interpretations of scientists are. These interpretations are based on a different worldview and do not come directly from science.

There are other worldviews as well. For example, pantheism suggests that all of creation is part of the divine. (This is reflected in some new-age philosophies.) In each worldview, there is a specific history taught that answers the basic question related to our identity. This helps us understand why the history presented in Genesis is so heavily attacked in our world today. Genesis forms the basis of the Christian worldview and is in conflict with the history proposed by other worldviews.

Building on the Solid Foundation

Some people today reject the idea that the Bible is a necessary foundation for everything. Many of these people believe that we came about through a series of chemical reactions, chance mutations, and natural selection. However, if this were true, then why would we have senses at all? Why would we expect our senses to give us accurate information about the world around us? Why would we expect to be able to understand the world around us? In fact, why would we trust any of our thoughts if they are merely the product of chemical reactions? Without the foundational belief in a biblical Creator, science does not have a strong logical foundation.

The Bible also provides the foundation for how we should live our lives. God often uses comparisons with the natural world to help us understand what He means. For example, trees can be very beautiful. Many trees produce fruit that we can eat. To accomplish this, a tree has roots that extend into the soil to provide it with support so it can remain upright. These roots also draw up water and nutrients from the soil. If a long drought occurs, the tree may not be able to absorb enough water through its roots. In this case, it may start to dry up and wither. In the following verses, God compares the person who studies and obeys the Bible to a tree whose roots are by a constant source of life-giving water.

Blessed is the man who does not walk in the counsel of the wicked or stand in the way of sinners or sit in the seat of mockers. But his delight is in the law of the Lord, and on his law he meditates day and night. He is like a tree planted by streams of water, which yields its fruit in season and whose leaf does not wither. Whatever he does prospers (Ps. 1:1–3 NIV).

MAKE A DIFFERENCE
ECO-FRIENDLY

Smart Shopping

When making choices in the grocery store, start checking labels for items that are in recycled or recyclable packaging and containers. They are often marked with the recycling symbol or indicate that it is made from recycled materials.

The Fruit of the Matter

Other analogies (comparisons) with plants occur in the parables of Jesus. In one case, Christ compared His followers to branches connected to a grapevine. He did this to emphasize the importance of us remaining in a healthy relationship with Him.

The vine supports the branches and provides the nutrients needed to produce the fruit. If the branch becomes separated from the vine, it will die. The fruit is the organ that protects the seeds and allows them to mature. The fruit from a good grapevine is also delicious and can provide nutrition to someone who eats it.

Those of us who continue to believe in Christ, trust His Word, and obey His teachings are remaining in Him. We are like branches truly connected to the source of life. We are loved, forgiven, and supported. We are infused (filled) with His life and strength so our life can bear fruit. Just like fruit from a good vine can be a blessing to someone who eats it, so fruit in our lives is a blessing to those around us. This fruit can be seen in the love we have for others (Gal. 5:22–23).

I am the vine, you are the branches. He who abides in Me, and I in him, bears much fruit; for without Me you can do nothing. If anyone does not abide in Me, he is cast out as a branch and is withered; and they gather them and throw them into the fire, and they are burned. If you abide in Me, and My words abide in you, you will ask what you desire, and it shall be done for you. By this My Father is glorified, that you bear much fruit; so you will be My disciples. (John 15:5–8).

Grafting in

New growth

Fruit

This is My commandment, that you love one another as I have loved you. Greater love has no one than this, than to lay down one's life for his friends (John 15:12–13).

This relationship also has implications (meaning) for properly understanding our world.

… that their hearts may be encouraged, being knit together in love, and attaining to all riches of the full assurance of understanding, to the knowledge of the mystery of God, both of the Father and of Christ, in whom are hidden all the treasures of wisdom and knowledge (Col. 2:2–3).

As we remain in Christ, we are intimately (closely) connected to the Creator, who has all wisdom and knowledge. Therefore, as Christians, our science can begin with God's perspective. Based on His Word, we are ready to study our world. The first thing we notice is that life on the planet is dependent on a mind-boggling complexity of relationships. This observation allows us to ask questions like: What are these relationships? How do they contribute to the survival of all creatures involved? Is human survival dependent on other creatures? Throughout His Word, God emphasizes relationship because relationship is a part of His character.

> No synonym for God is so perfect as Beauty.
> — John Muir

Chapter 10

> Level 1
> Level 2
> Level 3

Doing Ecology in God's Creation

We introduced you to the world of science in chapter 1, but did you know that scientists don't agree on the meaning of science? What they do agree on is that science is a human activity that strives to make meaning of this world. Throughout this book you have been *reading about* a few basic concepts ecologists are interested in. This final chapter will give you some suggestions about where to go from here in the *doing of ecology*. Before you begin, it is important that you start with the right foundation — God's Holy Word.

The apostle Paul, under the inspiration of the Holy Spirit, said that "for since the creation of the world God's invisible qualities — his eternal power and divine nature — have been clearly seen, being understood from what has been made, so that men are without excuse (Rom. 1:20). Have you ever stopped and wondered exactly which visible parts of His creation represent His invisible qualities?

MAKE A DIFFERENCE · ECO-FRIENDLY

Collecting Rain

Many people like collecting rainwater to use in watering gardens or other activities, but many states now have laws that restrict the collection of rainwater for personal use, especially those states in the western U.S. where water availability is often low yet consumption remains steady. Collecting rain means less is taken from private wells or city water systems, but large-scale collection is not feasible because rain is an important part of the water cycle in cleaning and replenishing streams.

If your area allows it, leave out a large bowl or bucket during the next rain, and see how much water you collect. Then use it to water flowers!

Explore with a Biblical Worldview

Go outside and find a location where you can sit and enjoy God's creation (e.g., back yard, park, forest, meadow, desert scrub, pond, lake, stream, or mountaintop).

Materials to bring would include the right clothing for the weather conditions, day pack, Bible, *The Ecology Book*, clipboard, drawing paper, notebook, and pencil. [Optional materials include paints (paintbrushes), crayons, colored pencils, magnifying glass (hand lens), and/or camera.] Once you arrive at your location, consider the following qualities of God (Romans 1:20):

• God is Life.
• God is Beauty.
• God is Provider (God is Love).
• God is Sustainer.

God is Life

Scripture is full of passages about the "living God" (Deut. 5:26; Josh. 3:10; Jer. 10:10; John 6:69; Acts 14:15; 1 Thess. 1:9; etc.). Ecologists study the relationships of living things with each other and their nonliving environment. What is life? Did you know that when something dies, there is no measureable difference in the chemical makeup of the creature before death and right after?

As you reflect on all of the organisms near you, remember that from God's perspective different creatures seem to have different types of life. If that is the case, and animals and humans both have "soul life," what makes human life different (see Gen. 1:26–27)? We would suggest that when ecologists study living things, we do not actually study life, but only the organisms that possess it.

God is Beauty

Many Scriptures reveal that God is beautiful (1Chron. 16:29; Ps. 72:19; Isa. 6:3; Ezek. 10:4; Rev. 18:1; etc.]. God's beauty is deep and whole and includes qualities like His love, grace, morality, and brilliant light. Look and notice the beauty around you. Take pictures, draw, paint, write poems or songs. These will help you tune in to the beauty around you. May the God of infinite beauty remind you of His beauty through His physical creation.

God is Provider [God is Love]

God provides through His love (Ps. 17:7, 25:6, 36:7; Isa. 63:7; 2 Cor. 13:11, 1 John 4:7, etc.). Not only does He provide for our needs, He also provides for His creatures. Look around you and consider His provisions. Consider the air and water that all creatures need. Think about the soil and the way its texture, chemistry, and nutrients are required for all life. Think of the sun that is, in part, responsible for weather and climate by moving winds, generating ocean currents, cycling water, and producing the energy that plants need to make sugars. The sun is just the right distance away, just the right size, and just the right color to provide the needed light and heat all organisms depend on. No wonder some have called earth the Goldilocks Planet; everything is just right. May you think about the air you breathe, the water you drink, and the food you eat and realize that our awesome Provider God created the heavens and the earth so that all would be provided for.

God is Sustainer

Scriptures such as Genesis 22:14, Matthew 6:25–33, and Hebrews 1:3 teach that God not only takes care of His creation, He sustains it. Remember the Nitrogen Cycle discussed in chapter 7? Chemical elements such as nitrogen, carbon, hydrogen, phosphorous, and molecules such as water are continually cycled by complex processes that require a whole bunch of organisms and specially created things to be in place and working together.

The Essence of Life

Did you know that few science textbooks ever define life, they just talk about life's characteristics? Biologists are in this strange position of not really agreeing about what life is. Is the essence of life physical or nonphysical (John 4:24; Gen. 2:7, Job 14:10; Acts 12:23)? Did you notice that when God created animals in Genesis 1:20–21, 24–25, He gave them the same "soul life" (Hebrew, *nephesh hayim*) that He gave to Adam in 2:7? Plants were not given that same life (Gen. 1:29–30).

Bottled Water

Bottled water takes many times more energy to produce, though there is very little difference between it and tap water from a faucet in your home.

Consider this from an environmental working group that studies how people impact the environment: "Every 27 hours Americans consume enough bottled water to circle the entire equator with plastic bottles stacked end to end. In just a single week, those bottles would stretch more than halfway to the moon — 155,400 miles."

Make a difference by reusing water bottles and recycling plastic bottles when you can. A few minutes of time can make a big difference and helps use our resources much more wisely!

Created for a Purpose

The special components of creation designed to sustain life include the chemical elements and their unique properties; the forces needed to put them together and move them; the sun; earth's rotation on its axis; the earth's 23.5 degree tilt; the earth's revolution around the sun; and many more.

Naturalists have made these observations too, and many have even admitted that this planet and universe look as if they were designed with man in mind. This principle has been named the Anthropic Principle, and though naturalists reject that idea outright, these observations are most consistent with the idea that they were designed by a Creator who desires to sustain His creation. God has created these biogeochemical cycles in order to make nutrients needed at the right time, in the right place, and in the right amounts throughout the organism's lifetime.

Studying Plants, Flowers, Animals, and Insects

Since the species is the basic unit of study for an ecologist, the next step is to get to know who lives in the place you have chosen. To do this, pick a group of creatures you are interested in and identify them. Are you interested in frogs? Turtles? Mammals? Trees? Insects? Just like anything else, you can't care about God's creation until you get to know them. Using a good field guide will help you to identify them. There are field guides on butterflies, general insects, amphibians (frogs and salamanders), reptiles (snakes, turtles, and lizards), mammals (bears, wolves, mice, bats), birds, fish, wildflowers ("weeds"), trees, snails, clams, and spiders. Just about any category of creature you might be interested in has a field guide you can learn from. You can even get field guides to the sounds of creatures like frogs, birds, crickets, and mammals. For those of you ready to take the night hike challenge, it will be fun to impress your friends with the animals you can identify, simply by knowing their calls.

Possible activities you can do include the following:

1. After identifying an organism, learn its scientific name and then start a "life list" of organisms you have identified.
2. After naming them, do a research project in the format of a poster, painting, PowerPoint, photo collage, crayon rubbings, or research report. Then report on the creature habitat, behavior, niche, range, and stewardship.
3. Briefly take in a species of organism and research their needs. Design an aquarium/terrarium habitat for this creature and take care of it for a brief time, then release it.

[Caution: make sure you choose common creatures that are not threatened or protected and check federal, state, and local laws to make sure there are no laws in place that make it illegal to take creatures out of their habitat without the required training and licenses. Many stream insects, reptiles, amphibians, and flowers fall in the protected category and should not be picked or taken from the wild.]

After you get to know the species, you can use some of the techniques in the appendix to study populations, communities, and ecosystems. Also, you can get involved with some of the organizations we listed in the various chapters and see if they can help you to design studies that will help you learn more about creation and help them monitor important creatures that they are concerned about.

Stewardship in Action

Remember that all of creation is God's and He has put us in charge of it. To be a good steward means to take care of His creation like He would. Things you can do include the following:

• Use the resources you have been given wisely, so that you may help others.

• Get together with your family, friends, and church to encourage one another not to litter and dump chemicals (like gasoline and antifreeze) into the environment, and to work together to do a trash cleanup in your neighborhood and community.

• Grow things and tend them. You could grow gardens at church and in your community and use your talents as image bearers of God to increase the beauty in your area.

• Grow community gardens. In this way, you not only learn the ecological growth needs of the food plants that people eat, but you are serving the community as a witness for Christ as well.

Field Guides and Resources

There are a number of good secular field guides available to students today, though these will not present a biblical worldview. The resources still can be used, but it is important to be discerning! Statements about the age of the earth and the evolution of animals are key concepts you need to watch for and address with your biblical worldview perspective when they arise.

Please be sure to check out the resource page in the back of this book to find educational and exciting books and DVDs that present important scientific details and biblical truths from a solid Christian perspective.

The Step-by-Step Recycling Process

Recycling includes collecting recyclable materials that would otherwise be considered waste, sorting and processing recyclables into raw materials such as fibers, manufacturing raw materials into new products, and purchasing recycled products.

Step 1. Collection

Collecting recyclables varies from community to community, but there are four primary methods: curbside, drop-off centers, buy-back centers, and deposit/refund programs. This is where you come in. Your part gets the whole process started as you collect paper, plastic, glass, and more around your home!

Step 2. Processing

Regardless of the method used to collect the recyclables, the next leg of their journey is usually the same. Recyclables are sent to a materials recovery facility to be sorted and prepared into marketable commodities for manufacturing.

Step 3. Manufacturing

Once cleaned and separated, the recyclables are ready to undergo the second part of the recycling loop.

More and more of today's products are being manufactured with total or partial recycled content. Common household items that contain recycled materials include newspapers and paper towels; aluminum, plastic, and glass soft drink containers; steel cans; and plastic laundry detergent bottles.

Recycled materials also are used in innovative applications such as recovered glass in roadway asphalt (glassphalt) or recovered plastic in carpeting, park benches, and pedestrian bridges.

Step 4. Purchasing Recycled Products

Purchasing recycled products completes the recycling loop. By "buying recycled," governments, as well as businesses and individual consumers, each play an important role in making the recycling process a success.

(from "Recycling: Steps to Recycling a Product", http://www.epa.gov/recycle/recycle.htm)

Made From
100% Recycled Materials

4

4

3

3

1

1

2

2

WE
RECYCLE

Tips for Doing Good Science Experiments

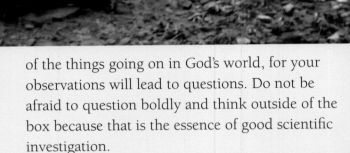

By building upon the foundation of God's Word, biblical creationists can study the world God has given us. Over time, scientists have developed ways of studying our world so that mistakes and improper thinking can be decreased. This way the conclusions we make can be as accurate as possible. Keep in mind that because people are not perfect, no experiment is perfect either.

There are many methods or ways of doing science. One of the most common methods used is the empirical method (operational science) in which you design an experiment in order to answer a question you have asked. In this book, you will have many opportunities to design your own experiments by using the empirical method so you can learn a little more about God's world. Here are a few tips that we think will help you both enjoy the scientific process and do it well.

1. The best tool that you bring to the scientific process is your brain. Your brain is not only a wonderful gift from God, but it also has been described as the most complicated physical substance in the universe. With your brain, you can make good observations and think clearly.

2. Scientific discoveries are often birthed by those who are using their brains to observe things carefully, so keep your eyes open. Become consciously aware of the things going on in God's world, for your observations will lead to questions. Do not be afraid to question boldly and think outside of the box because that is the essence of good scientific investigation.

3. From the questions being asked, hypotheses can be stated. A hypothesis is an educated guess written as a statement that can be tested by experiment. Scientists test two hypotheses at once: the alternative and the null. An alternative hypothesis, or research hypothesis, is what you are interested in testing. You are looking for patterns or relationships. The null hypothesis predicts no patterns or relationships. For example, while visiting a big city, you might observe that there are no lichens growing on trees. One question that can be asked is: Why aren't there any lichens growing on trees? There are many reasons why lichens may not be growing on trees, but one testable alternative hypothesis symbolized as (HA) may be: Sulfur dioxide (SO_2) pollution is affecting lichen growth. In other words, you think that there is a relationship between the pollutant SO_2 and lichen growth. The null hypothesis, symbolized

as Ho, may be stated this way: SO_2 pollution does not affect lichen growth. In other words, there is no relationship between SO_2 and lichen growth. Both of these statements can be tested at the same time. Let's say you noticed a possible relationship between SO_2 and lichen growth on trees. Your measurements suggest that the greater the SO_2, the less the lichen growth. Therefore, you can say that your measurements favor a relationship between sulfur dioxide and lichen growth (the alternative hypothesis), and therefore, you can reject the null hypothesis.

4. Once you have made an observation, asked a question, and stated the alternative and null hypotheses, you are ready to test the hypotheses. Testing hypotheses means you are doing an experiment. Here are a few things to keep in mind:

a. Variables are measurements or things that may affect what you are interested in testing. They tend to be changeable. When experimenting, change only one variable at a time and try to keep everything else the same.

b. The variable you change is called the independent variable. In our example above, you are interested in how the amount of sulfur dioxide affects lichen growth. In this case, the amount of sulfur dioxide is the independent variable.

c. A dependent variable (responding variable) is the measurement you think is affected by changing the independent variable. Can you identify the dependent variable in our above example? Good work! Yes, it is lichen growth on trees. Why? Because you are testing to see if changing the amount of SO_2 affects how lichens grow. Stated differently, is lichen growth dependent on the amount of SO_2 in the air?

d. Word your questions and hypotheses so the independent and dependent variables are in them. Our example question: How does the amount of SO_2 affect lichen growth on trees? Our example alternative hypothesis (HA): The amount of SO_2 affects lichen growth on trees.

e. You always want to have a control when you are designing an experiment. A control allows you to make comparisons. In this case, your control would be to observe lichen growth with no sulfur dioxide present. That way you can compare lichen growth with zero SO_2 present and see if lichen growth changes with increasing amounts of SO_2.

f. As much as possible, all other variables should be kept the same. When doing ecological research, it is impossible to control all variables. In our example, a few variables we would try to control include studying the same lichen species and studying in the same general climate, region, elevation above sea level, time of year, and time of day.

g. Later you'll want to graph your data. The independent variable goes on the x axis, and the dependent variable goes on the y axis.

h. Avoid bias. Bias means that you are favoring certain outcomes, whether you know it or not. For example, if you were just sampling lichens in a city and not away from the city, you are favoring lichen growth in cities, without comparing the data with lichen growth outside of the city. Biasing data can cause you to draw the wrong conclusions.

For further information see http://www.suite101.com/content/experimental-design-in-science-investigations-a146689.

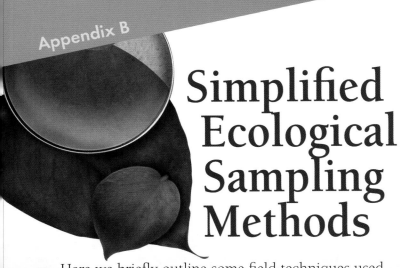

Simplified Ecological Sampling Methods

Here we briefly outline some field techniques used by ecologists that might give you some ideas for your own field experiments. This by no means includes all techniques, and each technique can be tweaked according to the needs of the ecologist.

First of all, it is possible that you will be handling some of these organisms. Here are a few things that will help you be better stewards of God's creatures:

1. Many places require a permit for catching and identifying creatures. Small mammals and stream insects are two examples. You will need to research your state's (or country's) guidelines about catching animals.

2. Many plants are protected for various reasons. It is always a good idea, when you are observing plants, to view them in their natural state and not pick them out of the ground. A good field guide to protected plants would be a handy resource.

3. If you are catching amphibians, like frogs and salamanders, the oils in your hands may hurt them because they have sensitive skin. So before handling, get your hands dirty or wet from the natural materials around you.

4. If you are turning over rocks and logs while looking for critters, remember that this is their habitat and sometimes it takes many years for these objects to become good homes. It is always a good idea to carefully roll the rock or log toward you, look to see who is home, and carefully replace it the way it was. You move it toward you because you will have a barrier between you and the creatures underneath. This is important in case there is a venomous snake underneath it.

5. Wash your hands after handling all animals.

Any time you do field experiments and observations, you should record, at the very least, the following information in a field notebook: location, time, date, researchers, air temperature (taken in the shade), wind direction/speed, and sky conditions (i.e., cloudy, partly cloudy, sunny). The reason that data are written is because it is easy to forget details of a particular day in the field. This way you can always go back to your notes and think about how your measurements might reveal things you never thought about before.

Ecologists often have to sample populations of organisms. Sampling is done because there is no way to determine exact numbers of critters in a particular habitat. Sampling techniques also depend on the ecological assumptions one makes. Different techniques all have their advantages and disadvantages, depending on the information needed.

When you want to sample various parts of a habitat, you want to take many random samples in places that represent the habitat as a whole. Random

sampling helps to eliminate bias. (See "Avoid bias," Appendix A.) Another example of bias can occur when you are sampling stream creatures. If you were to look for just the big ones, you would be favoring the big creatures in the sample. When sampling stream creatures, you want to count and identify both the big and small creatures.

One way to control the size of your sample is to use a quadrat or plot.

Quadrat/Plot — The plot can be a circular hoop of a particular diameter or more commonly a square frame that is one meter long by one meter wide. (Most scientists of the world use the International System of measures [SI or metric system], and that would be good for you to learn.) Depending on the size of the area studied, quadrats/plots can be much larger or smaller. With these measured areas, you are ensured of controlling the size of the area that you sample, which keeps you consistent.

Systematic Sampling

A. Line transect — If you wanted to sample a specific type of animal, like some salamanders that live under forest debris or plant communities, you would randomly choose various directions and then lay out a line through the woods, along a stream, or in open fields with a measuring tape. You can then walk along that line and at specific points, identify the creature(s) you are looking for. Line transects are used to give you an idea about how communities are changing and who lives where but do not give you as much detailed information as a belt transect.

B. Belt transect — This is the same as above except at random points along the line you can put a quadrat down and sample everything in the quadrat. This survey can give more information on density of species, which is a measure of how many species there are over a certain area (e.g., six salamanders per square meter).

Visual Encounter Surveys — These can be done for sampling animals like frogs, birds, salamanders, and insects. Each person walks through an area and carefully searches for the creature you are interested in. Record the time, the number of people, and how many animals were seen. This gives you a general idea of who lives there.

There are many other techniques and ways to measure variables. And when you study the details of these methods, you will find that there is a great deal of math involved. Hopefully these few techniques and insights will give you a flavor as to the kinds of variables and measurements considered when ecologists investigate ecosystems.

Abiotic — the nonliving portion of an ecosystem. Examples include water, air, and rocks … 10, 12, 16, 24, 26, 30

Anthropic principle — Principle that Earth and the universe seem designed with man in mind … 77, 79

Arbuscular mycorrhizae — fungal symbionts of plants that form branching structures from their hyphae called arbuscles. These arbuscles form in little compartments within root cells and are the location of nutrient transfer between the fungus and plant … 42, 45, 50

Arbuscles — branching hyphal extensions of an arbuscular mycorrhiza that penetrate a root cell for the purposes of nutrient transfer between a fungus and a plant … 45

Asexual reproduction — a way of making new individuals without sexual relations between a male and female … 32, 35

Autotroph — a producer organism that obtains its carbon from carbon dioxide … 27–28

Baraminology — a uniquely creationist study of the biodiversity of organisms and how they are grouped from a biblical perspective. The word baraminology comes from the Hebrew: *bara* [to create] and *mín* [kind] and means the study of the created kinds … 10, 12

Biogeochemical cycles — Cycles created by God to make nutrients available at the right time, right place, and in just the right amounts … 77, 79

Bioindicator — a living creature that can be used to determine the health of an environment and how the environment changes over time … 32

Biological diversity (biodiversity) — the number and variety of organisms and habitats in an area (also includes the genetic differences between creatures in an environment) … 10, 14, 17, 24, 26

Biomatrix — a creation biology term that refers to organisms, like nitrogen-fixing bacteria, that bridge the gap between the living and nonliving — also called an organosubstrate … 52, 57

Biome — a very large ecosystem — examples include the tropical rain forest, desert, and eastern deciduous forest … 10, 14

Biomonitoring — the process of studying areas where bioindicators live, and over time, measuring how changing environmental conditions affect them … 32, 39

Bioremediation — the technology that uses organisms to clean polluted environments … 42, 50

Biosphere — the living portion of planet Earth … 10, 14, 17

Biotic — the living portion of an ecosystem. Examples include bacteria, animals, plants, protists, and fungi … 10, 12, 14, 24, 27, 30

Botanist — a scientist who studies plants … 42–43

Carnivore — consumers that kill and eat other consumers … 24, 27

Cecotrope — a nutrient-rich soft pellet passed out of an animal that is re-eaten and provides the animal with important B vitamins and amino acids … 67

Cecum — a pouch-like structure connected between the small intestine and colon or large intestine where important digestive microbes live. Found in some animals … 67–68

Chemotroph — a producer organism that obtains its energy for life from chemicals such as sulphates in the surrounding environment … 27

Chemoautotroph — a producer organism that obtains its energy from chemicals in the environment and carbon from carbon dioxide. (Example: deep ocean bacteria) … 27–28

Chemoorganotroph (Heterotroph) — a consumer organism that obtains its energy and carbon from the food that it eats … 27–28

Citizen science — a way for ordinary people to take part in real science projects being done by scientists … 18, 23

Commensalism — a type of symbiosis between two or more different organisms where one is benefited and the other is not affected one way or the other … 6–7

Consumer (see chemoorganotroph or heterotroph)

Condensation — the process by which water changes from a gas to a liquid … 10, 17

Cyanobacteria — a group of photoautotrophic bacteria … 28, 33, 36, 49

Decomposer — organisms that break down chemical compounds into important nutrients and make them available for other organisms in the ecosystem. They are God's recyclers … 24, 27

Diazotrophs — bacteria that have special enzyme systems that allow them to directly use nitrogen gas and make it available to plants. *Azo* comes from the French word for nitrogen (*azote*) … 54

Dictyosome — the flattened sacs in a Golgi apparatus organelle, especially in plants, that are involved with marking, shipping, and transporting substances in a cell … 56

Dominion mandate — God's command for the humans He created to take care of and rule over the earth and all the other creatures of the planet … 3, 18

Ecological pyramid — a model used by ecologists to show how energy moves through an ecosystem … 24, 28

Ecological system (ecosystem) — location where the relationships of organisms with each other and their environment takes place. It can be as small as a drop of water to as large as earth's biosphere … 10, 14, 16, 19, 23, 25–30, 33, 47, 51

Ecology — the study of the relationships of living things with each other and with their nonliving environment … 5, 10, 12–14, 18, 23–26, 28, 30, 41, 76

Ectomycorrhiza (plural **Ectomycorrhizae**) — fungal symbionts of plants that generally form on the outside of roots by making a sheath or covering over the root … 42, 44–46, 49

Embryo — a creature in the early stages of development, before birth … 13, 49

Endomycorrhiza (plural — **Endomycorrhizae**) — fungal symbionts of plants that actually penetrate the cell wall of root cells … 42, 44–45, 49

Endosymbiosis — a symbiosis in which one organism lives inside the body of another and both work together as a single organism … 52, 56

Evaporation — the process by which water changes from a liquid to a gas … 10

Evapotranspiration — part of the water cycle where plants lose their water to the atmosphere by the process of evaporation, mostly through the leaves … 10, 16

Fermentation — the process by which certain bacteria or fungi break down carbohydrates to produce an acid (as in yogurt) or an alcohol (as in winemaking) … 63–64, 66–67

Food chain — a simple model showing how energy moves through a simple feeding relationship. For example: plant → grasshopper → mouse → snake (grasshopper eats plant, mouse eats grasshopper, snake eats mouse). Arrows show the direction of energy … 24, 28–29

Food web — a model showing many interconnected food chains … 24, 28–29

Fragmentation — a type of asexual reproduction in lichens that happens when a little piece of lichen breaks off, gets blown away to a new location, and a new lichen grows from it … 32, 35

The inside of a ginkgo seed, showing the embryo

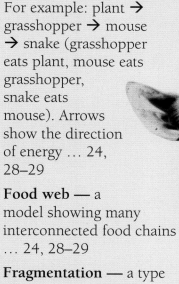

Giraffes are herbivores

This lichen is composed of both yellow and white parts. Black rhizines extend at the bottom, ready to serve to anchor the plant. A few developing moss sporophytes (green) are also visible.

Green Manure — nitrogen fixing plants grown on the land and then plowed under soon after they flower, in order to enrich farm soil with nutrients … 57

Gnotobiotic — the study of living things that have never been exposed to microorganisms except those selected by the researcher … 63, 68

Habitat — where an organism lives … 18, 24, 26, 33, 35, 70, 80, 86

Harmony — a pleasing and beneficial relationship … 3, 6, 8–9, 25

Herbivore — a consumer organism that eats only plants … 24, 27

Heterotroph (chemoorganotroph) — a consumer organism that obtains its carbon and energy from the food that it eats … 27–28

Heterotrophs (consumers) — organisms that must get their energy by eating other organisms … 27–28

Hybridization (hybrids) — when two different organisms (different species or genera) mate and have offspring. Examples include tiger/lion, camel/llama, false killer whale/bottlenose dolphin, polar bear/grizzly bear … 10, 12

Hydrologic cycle — the continuous movement of water above, on, and below the surface of the ground … 16

Hypha (plural **hyphae**) — a long, thin, and branching structure of most fungi. It may be involved with processes like digesting and/or absorbing food nutrients, asexual reproduction, or protection, depending on the fungus … 32, 35–36, 39, 42, 44–46, 48–49

Legume — plants like beans, peas, clover, and alfalfa that form small nodules (little balls) on their roots where nitrogen-fixing bacteria live and fix nitrogen for the plant to use … 52–53, 55–58, 60, 65

Lichen — an organism made of two or three organisms working as one. A lichen can be a relationship with a fungus and alga, fungus and cyanobacterium, or a fungus, alga, and cyanobacterium … 32–41, 84–85

Lichenologist — a scientist who studies lichens … 32

Microbiota — the microorganisms that normally live on the skin or inside the digestive system. Microorganisms are so tiny a microscope is needed to see them … 62, 64-69

Mutualism — a type of symbiosis where two or more organisms benefit from one another. Some researchers equate symbiosis and mutualism as the same relationship and others do not … 6–7

Mycelium (plural — **Mycelia**) — a large number of hyphae on a fungus … 32, 36, 46

Mycobiont — the fungal partner or symbiont of a lichen … 32, 36

Mycoheterotroph — a plant that does not photosynthesize and has to get its carbon nutrients from mycorrhizae, who must get them from photosynthesizing plants … 42, 46

Mycorrhiza (plural — **Mycorrhizae**) — a fungal symbiont of plants — literally means "fungus root" … 42–51

Nephesh Hayim — Hebrew words meaning "soul life;" the life given by God to animals and men only. Can be thought of as Biblical or soul life … 77, 79

Niche — the role or position the organism has in its environment. It can be defined as how the organism feeds, finds shelter, and reproduces. It can also be defined as what the creature needs in terms of living space, amount of water, temperature, and other environmental conditions … 24, 26, 80

Nitrogen fixation — the process done by bacteria or cyanobacteria in which nitrogen is combined with other elements so it is in a form that plants can use … 52, 57

Nutrient — a substance, like nitrogen, that provides nourishment for an organism … 28, 39, 46, 52, 57, 67

Omnivore — consumer organisms that eat both plants and animals … 24, 27

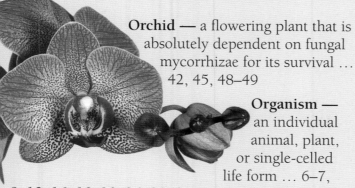

Orchid — a flowering plant that is absolutely dependent on fungal mycorrhizae for its survival … 42, 45, 48–49

Organism — an individual animal, plant, or single-celled life form … 6–7, 9, 13–16, 18–19, 26, 28, 33, 36, 40, 45, 47, 50, 68, 70–71, 78–80, 86

Organosubstrate — a creation biology term that refers to organisms, like nitrogen-fixing bacteria, that bridge the gap between the living and nonliving — also called a biomatrix … 52

Pantheism — suggests that all creation is part of the divine … 70, 72

Parasitism — a type of symbiosis of two or more organisms where one benefits from the other by taking nutrients from it, sometimes causing harm … 7

Pathogenic — an organism that causes disease … 62–63

Pelotons — fungal hyphae of orchid mycorrhizae found in orchid cells. They are short lasting and digested by the orchid to get nutrients provided by the fungus … 42, 48–49

Phloem — a type of vascular tissue where groups of cells are working together to transport sugars and other materials throughout a vascular plant … 42, 45

Photoautotroph — a producer organism that obtains its energy from light and carbon from carbon dioxide. (Examples: plants, algae, cyanobacteria) … 27–28

Photobiont — the photosynthetic partner or symbiont (alga or cyanobacterium) of a lichen … 32, 36

Photomorph — different forms of the same lichen depending on photobiont … 32, 36

Photosynthesis — the process by which producers (like plants) get their energy from light in order to build important carbon nutrients … 24, 28, 31, 36, 52

Phototroph — a producer organism that obtains its energy from light … 27

Phytoremediation — is a form of bioremediation where plants are used to take toxins out of soil, as in the case with arbuscular mycorrhizae … 42, 50

Population — the population concept is difficult to define. But it can be simply described as the group of one species in an area that have equal chances of mating with one another. (Note: This definition is limited and too simplistic to describe most situations in an ecosystem.) … 10, 13, 29, 64

Precipitation — Water falling from clouds in the form of rain, sleet, hail, or snow … 10, 17

Producers (autotrophs) — organisms able to produce organic compounds, like carbon nutrients, using light or chemical energy … 27–29, 47, 53

Protocol — a detailed plan for doing a science experiment … 10

Recharge — the process by which water moves down through the soil to the groundwater, which is the part of the soil that is soaked with water … 10

Recycled — an item that has been reprocessed for another use … 57, 70, 73, 82

Rhizine — strands of hyphae found on lower surface of many foliose lichen. Used for anchoring the lichen to what its grown on … 35, 91

Rhizobia — certain nitrogen-fixing bacteria that form a very close relationship with legume plants … 52–53, 55–57

Ruminant — an animal (such as deer, sheep, and cattle) that has a four-chambered stomach and chews its cud … 66

Saprophyte (saprobe) — a heterotrophic organism that grows on formerly living organisms and gets needed nutrients from them. May also be decomposers … 24, 27

Scavenger — a heterotroph that eats already dead organisms … 24

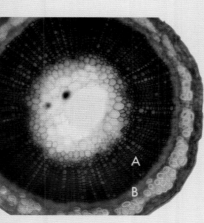

Cross-section of a vasuclar plant stem showing xylem (A) and phloem (B).

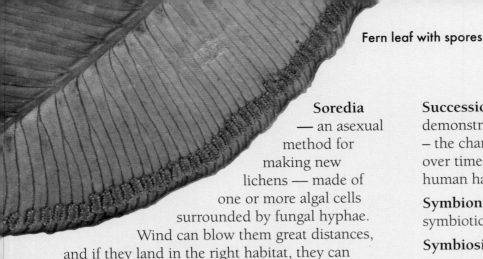

Fern leaf with spores

Soredia — an asexual method for making new lichens — made of one or more algal cells surrounded by fungal hyphae. Wind can blow them great distances, and if they land in the right habitat, they can produce new lichens … 32, 35

Species — the species concept is complicated and controversial, but most ecologists would agree that, at its basic level, a species is a group of creatures that have a stable structure and that can reproduce with each other. The offspring of these parents are then able to reproduce with each other … 10, 12–14, 36, 39, 46, 49–51, 56, 64, 80, 85

Spore — a small, asexual reproductive cell that can develop into a new individual … 32, 44

Stewardship — comes from the old English word *stigweard*, which means "guard of the hall." The word implies that a steward is responsible for taking care of something for someone else. In this case, humans have been given responsibility to take care of God's creation for Him … 14, 18–19, 20, 55, 61, 80–81

Sublimation — the process by which water changes from a solid directly into a gas, without going through the liquid phase … 10, 16

Succession (community) — a process that demonstrates God as Sustainer and Restorer of beauty – the changing of ecological communities of an area over time and after disturbances like flood, fire, and human habitat destruction … 42, 46

Symbiont — an organism that is a partner in a symbiotic relationship … 32, 36

Symbiosis — a long-term relationship between two or more organisms … 7, 32–33, 36, 44–45, 56, 70

Symbiosome — a specialized organelle within the host cell that houses microbes like bacteria … 56

Thallus — the body of a lichen … 32, 36

Trophic level — each step in a food chain or food web. For example, autotrophs make up the first trophic level in all ecosystems … 24, 28–29

Vascular plant — a plant that contains the vascular tissue of xylem and phloem … 42, 45

Watershed (catchment) — the land area that drains into a particular lake, river, or ocean … 10

Xylem — a type of vascular tissue where groups of cells are working together to transport water up vascular plants from the roots … 42, 45

Watershed

Additional Resources

Whether you are looking to take a special family trip or just want to extend your educational program, these suggested resources are all written and designed with a solid, biblical worldview. You will learn about the history, ecology, wildlife, and geology of the most scenic national parks while you nurture and strengthen your faith. The world is an amazing testament to the creativity and power of our Creator. Enjoy these natural wonders in an educationally fun way!

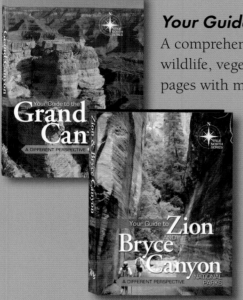

Your Guide to the Grand Canyon

A comprehensive guide with a biblical worldview! Discover the canyon's wildlife, vegetation, fossils, geology, and history. Includes 26 fold-out pages with maps, special overlook features, and more!

Your Guide to Zion and Bryce Canyon National Parks

This unique, full-color guide unveils God's powerful hand in the grandeur of the vertical walls of Zion, the colorful hoodoos of Bryce Canyon, and all the diversity of life found in these stunning displays of creation.

Your Guide to Yellowstone and Grand Teton National Parks

Within 188 pages, you will find travel tips, maps, details on the vast forests, grasslands, geysers, trails, flowers, hiking trails, wildlife, and more, vibrantly shown in hundreds of color photos in a complete guide from a biblical worldview!

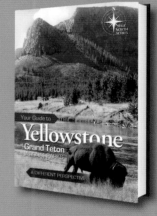

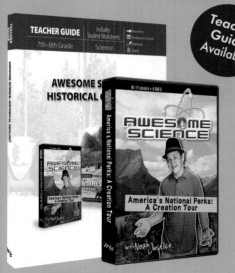

Teacher Guide Available

Awesome Science with Noah Justice

The *Awesome Science* video series takes teens and adults on a field trip around the world to explore geologic and historical evidence that supports the biblical record. Innovative, high-quality, and designed to make science fun, this new series for the whole family helps discover evidence that the Bible is the true history book of the world!

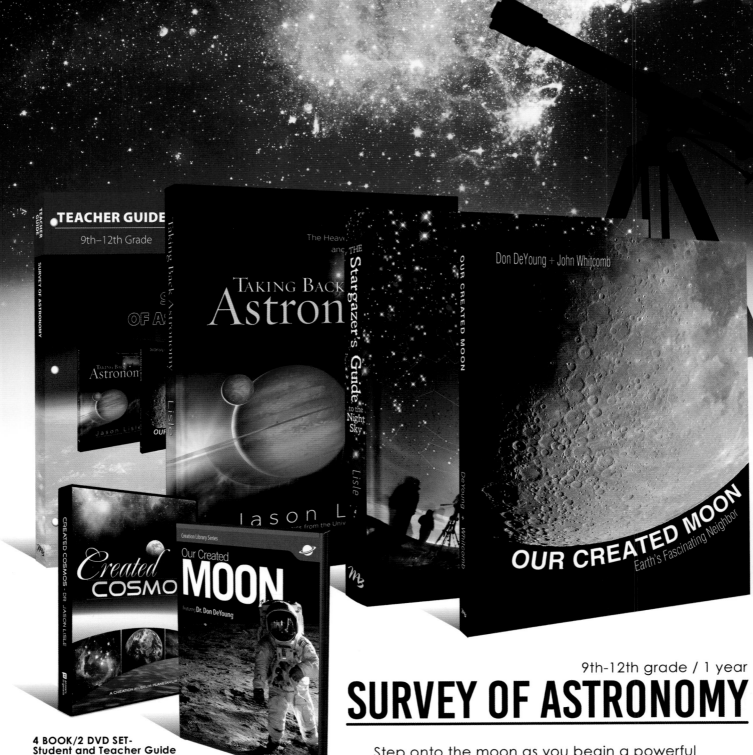

TEACHER GUIDE
9th–12th Grade

The Heav...
and...

TAKING BACK
Astron...
OF A...

THE
Stargazer's
Guide
to the
Night
Sky

OUR CREATED MOON

Don DeYoung + John Whitcomb

Jason L...

OUR CREATED MOON
Earth's Fascinating Neighbor

Created COSMO... · DR. JASON LISLE
A CREATION-MUSEUM PLANETARIUM...

Creation Library Series

Our Created
MOON
Featuring Dr. Don DeYoung

4 BOOK/2 DVD SET-
Student and Teacher Guide
978-0-89051-766-6

Our Award-Winning
Wonders of Creation Series

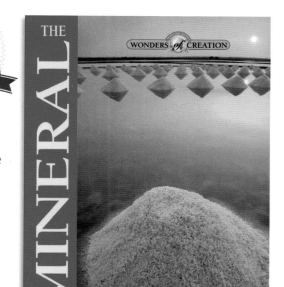
Filled with special features, every exciting title includes over 200 beautiful full-color photos and illustrations, practical hands-on learning experiments, charts, graphs, glossary, and index — it's no wonder these books have become one of our most requested series.

- **The Mineral Book*** reveals the first mention of minerals in the Bible and their value in culture and society.
- **The Ecology Book*** researches the relationship between living organisms and our place in God's wondrous creation.
- **The Archaeology Book*** uncovers ancient history from alphabets to ziggurats.
- **The Cave Book** digs deep into the hidden wonders beneath the surface.
- **The New Astronomy Book*** soars through the solar system separating myth from fact.
- **The Geology Book** provides a tour of the earth's crust pointing out the beauty and the scientific evidences for creation.
- **The Fossil Book** explains everything about fossils while also demonstrating the shortcomings of the evolutionary theory.
- **The New Ocean Book*** explores the depths of the ocean to find the mysteries of the deep.
- **The New Weather Book*** delves into all weather phenomena, including modern questions of supposed climate change.

*This title is color-coded with three educational levels in mind: 5th to 6th grades, 7th to 8th grades, and 9th through 11th grades.

8 1/2 x 11 • Casebound • 96 pages • Full-color interior
ISBN-13: 978-0-89051-802-1
JR. HIGH to HIGH SCHOOL

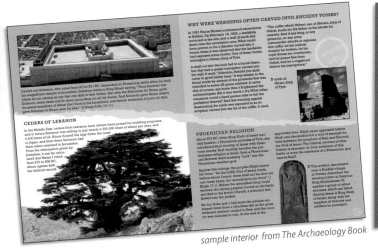

sample interior from The Archaeology Book

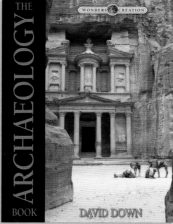

The Ecology Book
ISBN-13: 978-0-89051-701-7

The Archaeology Book
ISBN-13: 978-0-89051-573-0

The New Ocean Book
ISBN-13: 978-0-89051-905-9

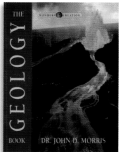

The Geology Book
ISBN-13: 978-0-89051-281-4

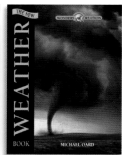

The New Weather Book
ISBN-13: 978-0-89051-861-8

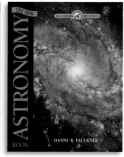

The New Astronomy Book
ISBN-13: 978-0-89051-834-2

The Fossil Book
ISBN-13: 978-0-89051-438-2

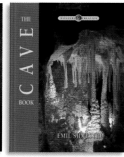

The Cave Book
ISBN-13: 978-0-89051-496-2

Daily Lesson Plan

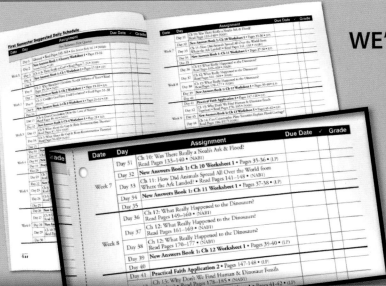

WE'VE DONE THE WORK FOR YOU!

PERFORATED & 3-HOLE PUNCHED
FLEXIBLE 180-DAY SCHEDULE
DAILY LIST OF ACTIVITIES
RECORD KEEPING

"THE TEACHER GUIDE MAKES THINGS SO MUCH EASIER AND TAKES THE GUESS WORK OUT OF IT FOR ME."

★ ★ ★ ★ ☆

HOMESCHOOL

Master Books® Homeschool Curriculum

Faith-Building Books & Resources
Parent-Friendly Lesson Plans
Biblically-Based Worldview
Affordably Priced

Master Books® is the leading publisher of books and resources based upon a Biblical worldview that points to God as our Creator.

Now the books you love, from the authors you trust like Ken Ham, Michael Farris, Tommy Mitchell, and many more are available as a homeschool curriculum.

MASTERBOOKS.COM
— *Where Faith Grows!* —